A Catalogue of the Library

of

SIR RICHARD BURTON, K.C.M.G.

held by the

Royal Anthropological Institute

edited by

B. J. Kirkpatrick

British Library Cataloguing in Publication Data

Royal Anthropological Institute
A catalogue of the library of Sir Richard Burton, K.C.M.G., held by the Royal Anthropological Institute.
1. Burton, *Sir* Richard, *b.* 1821 – Library
I. Title
II. Burton, *Sir* Richard *b. 1821*
III. Kirkpatrick, B. J.
017'.6 Z997.B/

ISBN 0 900632 13 5

Royal Anthropological Institute
56 Queen Anne Street, London W1M 9LA

1978

CHAPTER II.

THE ANCIENT COURSE OF THE INDUS. CANAL IRRIGATION UNDER THE NATIVE PRINCES, AND THE PRESENT RULE. SYSTEM OF TAXATION IN THE TIME OF THE AMEERS.

WITH respect to the geographical position of the Indus in remote ages, little remains to be said. The ~~different opinions~~ concerning its course in the days of Alexander, and the various arguments for and against the theory of its ancient channel having been to the eastward of the present bed, have been discussed ~~*usque ad nauseam*~~.

The natives of Sindh now enter, to a certain extent, into the spirit of the inquiry; and, like true Orientals, do their best to baffle investigation by the strange, ingenious, and complicated lies with which they meet it. At Hydrabad, an old man, when questioned upon the subject, positively assured ~~me~~ that in his father's time the Indus was fordable from the spot where the Entrenched Camp now stands, to Kotree, on the opposite bank of the river. The people[1] abound in stories and traditions, written as well as oral, about the shifting of their favourite stream; and are, besides, disposed to theorize upon the subject. Some, for instance, will

Page 35 from Burton's copy of Sindh (Catalogue:3), extensively annotated for a 2nd edition (never published).

CONTENTS

FOREWORD

When in 1955 the trust for the safe keeping of Sir Richard Burton's Library was handed on by the Royal Borough of Kensington to the Royal Anthropological Institute, the Council of the Institute resolved to set in hand the compilation of a full catalogue of the collection, which alone would permit it to be made freely and permanently available to scholars, and thus not to remain as a buried talent. Fortunately our Library, which had grown up since the foundation in 1842 of the Ethnological Society of London thanks to diverse enthusiasms of many amateurs until it was one of the best anthropological libraries in the world, had by then come under the controlling hand of a professional librarian, Miss B. J. Kirkpatrick, who was indeed already established as a kind of 'resident architect' watching and guiding its growth into an even more valuable arsenal for all anthropologists; added to this, she became known in her spare time as a distinguished bibliographer in the field of English twentieth-century literature. It was clear that the allotment of the task to her was the greatest contribution which the Institute could make in Burton's memory and in the discharge of its trust.

During two decades of mounting financial pressures on the Institute (as on all learned societies) it was able – largely thanks to the generous Fellows mentioned in Miss Kirkpatrick's Preface – to engage extra staff from time to time so that she might devote some part of her time to this Catalogue. She was able to complete the compilation before going to the Australian Institute of Aboriginal Studies at Canberra for a period in 1976.

It is our hope that this remarkable record will be of considerable value to writers and others who have occasion to study Burton, or indeed his times; certainly there can be few literary men of the nineteenth century, let alone founding fathers of anthropology, who have been so fortunate in the disposition of their literary chattels after their death. Now that the necessary work of cataloguing has been well done, the world of learning has not merely the option but a positive obligation to consult so valuable a source book. There are no doubt many points, major and minor, on which, with its help, any writer wishing to touch on the life and ideas of Richard Burton may, and therefore should, improve his text whether by verification or by serendipitous discovery here of facts which have hitherto lain unregarded. We hope that Jan Morris's excellent Introduction will help to alert the world of literature.

Isabel Burton has been justly censured as a burner of books and, what is far worse, of potential books or manuscripts (of which the Institute, now that the age permits, would have been glad to assist publication); however, it must be set to her account on the credit side that she saw to the survival of the herein documented remainder.

WILLIAM FAGG
Honorary Librarian and Archivist

INTRODUCTION

By their books, it might equally be said, shall ye know them! Except perhaps for their pets, nothing reveals men more clearly than their libraries, and no lapidary inscription tells a life's tale better, or at least more honestly, than the row of titles on a domestic bookshelf, gathered there like wrinkles down the maturing years.

Few celebrated men can be more justly reflected in their reading-matter than is Richard Francis Burton in the library so meticulously recorded in this Catalogue. It is an astonishingly eclectic list of books, if only because Burton himself fell into no category. He was not exactly an intellectual, nor primarily a soldier, nor just an explorer, nor simply a scholar. He was an iconoclast in some ways, a traditionalist in others. He was an individualist on an alarming scale, suffering fools with difficulty, but his childless marriage to the devoted fatuous Isabel Arundell lasted forty years. He was a man of Satanic appearance and hardly less demoniac reputation, yet he spent his last years in the respectable minor office of British Consul at Trieste, and did not hesitate to accept the honour when his services to the State were recognized with a belated knighthood.

Burton would have been a striking man in any age, but part of his fascination arises from his historical setting. Despite appearances and perhaps intentions he was pre-eminently a child of his times – High Victorian times, that is, when the British Empire was approaching its apogee, and British enterprise was at its most furious and inventive. Though he delighted in flouting Victorian shibboleths, and claimed that England was the only country where he never felt at home, Burton was nevertheless a Great Victorian himself, sharing far more of the prime Victorian qualities than he perhaps recognized (or admitted).

It is true that he was hardly a success in the Oxford of the 1840s, where he felt himself to have fallen among grocers, but his restless and insatiable energy, his determination to look beyond the horizon whether physical or figurative, were altogether of his time and people. It was the British Empire, that grand employment agency for the maverick, that gave him the world's freedom, and it was the Victorian instinct for synthesized exactitude that we can see in his passion for detail and comparison.

Like his century, he was at once rationalist and romantic, and it is no surprise to find in the same library shelves both Nos 821 *(The Book of Noodles)* and 2113 (a condensed speech by the Vice-President of the Slavonic Committee of Moscow in 1876). If he was a serendipitist of marvellous facility, he was also a thoroughly painstaking scholar. This is a pleasure library, but it is an efficient working library too, as we may see from the many editions it contains of the poems of Camoens and *The Arabian Nights* – guides, indices and examples for Burton's own translations of those works.

Burton's flamboyant taste for adventure was, of course, thoroughly contemporary. His was one of the great ages of adventure, when the drive to reveal the last unknown places of the earth proceeded in parallel with the mighty development of steam. Burton was fortunate not just in his moment, but in his nationality. Britain was the greatest Muslim and the greatest African Power, and it was part of the truth of the times that this irrepressible Briton should be among the earliest Christian visitors to Mecca, the first European to return unscathed from Harrar, and a member, however bitter and contrary, of the company of explorers who between them solved the mystery of the Nile sources.

Nor was he so eccentric or anachronistic as is sometimes suggested in his indefatigable quest for anthropological and ethnological truths. There were thousands like him in the service of the British Empire. It was his own Commander-in-Chief, General C. J. Napier, who commissioned Burton to investigate the pederast brothels of Karachi, so firing his life-long interest in oriental erotica, and all over the Empire there were servants of the Crown just as eagerly investigating unfamiliar customs, languages and literatures. Burton was exceptional chiefly, genius apart, in his refusal to conceal findings, or to compromise scholarship, in the face of conventional morality.

One would hardly have expected patronage of the first unexpurgated edition of *The Thousand and One Nights* to be among the Victorian imperial achievements, but in a very real sense the Empire was the patron of Burton's masterpiece. He was a lifelong employee of the State himself, and it was by courtesy of the royal firman, over many years, that he was enabled as the summit of his life's work to present the magnificent old stories Nos 91–4 in their full splendour to the wondering West.

But then just as in its highest aspirations the idea of Empire transcended faith or nationality, making the Great Queen almost a universal Mother-Figure, so Burton too was really a man beyond racial or religious limit. He was a true mystic, seeing always beyond the detail, fascinating though that was, to grander realities beyond. It has been suggested that he was actually an initiated member of a secret Sufi order: certainly in his confessional poem *The Kasidah* his voice is authentically Sufic, and I believe that the great mass of his own work, so varied and apparently contradictory, really adds up to a view of all life as a grand unity.

It was like a pointillist view, in which each infinitesimal segment of the picture contributes to a scene of very different scale and meaning. Every library, of course, has such a character, but by the nature of its owner Burton's has more than most. Though most of his private journals were destroyed at his death by Lady Burton – "Let the world rain fire and brimstone on me!" – still the fascinating range of this not very large collection suggests unmistakably the power of the personality that assembled it. For all his faults and even fooleries, Burton was a majestic man: and there is majesty to his library too, as though his brooding prescence roams the shelves and titles still, making them all one whole.

JAN MORRIS

PREFACE

Sir Richard Burton (1821–90) was associated with the Ethnological Society and its successors from 1861 until his death. A founder and active member of the seceding Anthropological Society, he took the chair at the inaugural meeting on 6 January 1863, became its first Vice-President, and its President in 1867. He was a member of the Council of the Anthropological Institute from 1872–3, and died in office as Vice-President.

Burton made his well-known remark 'the rooms of the Anthropological Society now offer a refuge to destitute truth' at the public dinner given by the Society on 4 April 1865 to mark both his departure for the Santos Consulate, Brazil, and the election of five hundred Fellows since the first meeting of eleven. He appears to have made a somewhat similar remark during the discussion following his paper 'Notes on certain matters connected with the Dahoman' given to the Society on 1 November 1864. Though associated with the seceding body, the Anthropological Society, Burton is reported to have said on its reunion in 1871 with the Ethnological Society as the Anthropological Institute, 'the two Societies always should have been one.' However, in 1873 he was again among the secessionists of the Anthropological Institute who formed the London Anthropological Society of which he became one of its Vice-Presidents. In a letter to the Director, Sir Edward Brabrook, of the Anthropological Institute he gave his reason as 'the deadly shade of respectability, the trail of the slow-worm, is over them all.' Burton published two papers in *Anthropologia*, the organ of the London Anthropological Society, which is not held by the RAI Library. An offprint of one is in his Library (*see* 45*b*).

Burton contributed many papers and other communications to the journals of the two Societies and the Institute. His first was to the *Transactions of the Ethnological Society* in 1861 (*see* 12); extracts from his last communication to the Institute 'On the Akkas' were read and commented on by Sir Edward Tylor on 27 March 1888 (the title only is noted in the *Journal*, Vol. 18, p. 121).

The three most recent lives of Burton are:

Brodie, Fawn M. *The devil drives: a life of Sir Richard Burton.* London, Eyre & Spottiswoode, 1967

Farwell, Byron. *Burton: a biography of Sir Richard Burton.* London, Longmans, 1963

Hastings, Michael. Sir Richard Burton, a biography. London, Hodder & Stoughton, 1978. For an account of the conflict between the Ethnological and Anthropological Societies reference should be made to 'What's in a name? The origins of the Royal Anthropological Institute (1837–71)' by George W. Stocking, Jr, *Man*, 1971, N.S. Vol. 6, pp. 369–90.

N. M. Penzer gives a description of Burton's collection in his *An annotated bibliography of Sir Richard Burton, K.C.M.G.*, London, pp. 291–8 which is a valuable source of information.

Notes on the catalogue:

1. The RAI was able to accept Sir Richard Burton's Library from the Royal Borough of Kensington in 1955 through the generosity of several Fellows of the Institute, Dr G. Caton-Thompson and Dr B. G. Campbell, the late Mrs B. Z. Seligman, the late Lord Sieff and the late Dr J. C. Trevor, and the United Africa Company who made it financially possible for the RAI to house it.
2. *Grants.* Since the initial grants, three have been received:
   (a) *Bollingen Foundation Inc.*, New York. Many of the books were received in very poor condition due to flooding. As a result of the grant all such books were either

restored to their original condition as far as possible, or, the original binding copied where its state made retention impossible.

(b) *The Leverhulme Trust Fund.* A very welcome grant was received towards cataloguing the collection in the initial stages.

(c) *Esperanza Trust for Anthropological Research.* Funds were made available to free the Librarian from some routine library work so that the catalogue could be completed. Without this generous grant, the delay in the publication of the catalogue would have been even greater.

3. *Arrangement of the catalogue.* The entries are arranged under broad subject headings which are related to Burton's own books, his interests, and his residences. His own works, Items 1–108 are arranged chronologically; all other entries are alphabetically by author under each subject heading.
4. *Index.* This includes the names of editors, translators, etc., correspondents, and donors of books. Authors' names, and more specific subject headings are only included either where association with another work is necessary, or to bring together books on a subject of particular interest to Burton such as the entry under Háfiz.
5. *Books or pamphlets shelved out of sequence.*
   (a) An item number within square brackets at the end of an entry indicates that the work in question is either inserted or bound in the volume cited.
   (b) p., q., cabinet, indicate pamphlet and quarto sequences, and the map cabinet.
6. *Abbreviations used and the use of terms.*
   (a) Penzer – *An annotated bibliography of Sir Richard Francis Burton, K.C.M.G.*, by N. M. Penzer. London, A. M. Philpot, 1923; reprinted by Dawsons of Pall Mall, London, 1967.
   (b) *The Life* – *The life of Captain Sir Richd. F. Burton, K.C.M.G., F.R.G.S.*, by Isabel Burton. London, Chapman and Hall, [1893].
   (c) IB – Lady Burton.
   (d) holograph and manuscript notes – holograph is used for leaves or notes in Burton's hand, and manuscript notes for other hands.
7. *Indications of Burton's use of a book.*
   Entries which have the following lower case letters beside the item number carry the following indications of association with Burton:

a Burton's signature or initials: Richd F. Burton; Richard F. Burton; R. F. Burton; R.F.B.
b book-plate; white, 3³⁄₁₀ × 2⁷⁄₁₀ in.: [*illustration of a talbot's head erased,* ⁷⁄₁₀ × ⁷⁄₁₀ in.] / [*in black letter:*] Richard F. Burton.
c visiting card; white, 1½ × 3 in.: *Captain Sir Richard F. Burton. / Athenaeum.*
d annotations and marked passages (the latter include passages marked N.B.).
e marked passages only.
f holograph sketches.
g holograph leaves inserted.
h newspaper, journal, clippings and other printed matter inserted; and manuscript material not by Burton.
i presented to Burton by the author unless specified otherwise, or carrying an autograph signature; date noted if given.
j autograph letters to Burton or Lady Burton inserted; sender noted, and date if given.
k from the libraries of Henry Raymond Arundell, Eliza Arundell, and other members of Lady Burton's family, and from Lady Burton's own collection.
l annotations and marked passages by another hand, many probably by Lady Burton

8. The following by Burton are not in the collection:
   1860. 'The lake regions of central equatorial Africa, with notices of the lunar mountains.' – offprint from the *Journal of the Royal Geographical Society*, Vol. 29 (*see* Penzer, pp. 67–8)
   1873. *The lands of Cazembe: Lacerda's Journey to Cazembe in 1798* – not in collection though recorded as present by Penzer (*see* p. 90)
   [1880] *The Kasîdah (couplets) of Hâjî Abdûx el-Yezdî . . .* translated and annotated by F. B. (*see* Penzer, pp. 97–103)
   1890. *Priapeia: or the sportive epigrams of divers poets on Priapus* (*see* Penzer pp. 150–3)
   1891. *Marocco and the Moors: being an account of travels . . .* by Arthur Leared second edition revised and edited by Sir Richard Burton (*see* Penzer, pp. 154–5)
   1898. *The Jew, the Gypsy and El Islam* (*see* Penzer, p. 158)

1901. *Wanderings in three continents* (*see* Penzer, pp. 158–9)
and the *Kama Shastra Society publications:*
1883. *Kama Sutra of Vatsyayana* (*see* Penzer, pp. 163–71)
1885. *Ananga-Ranga (stage of the bodiless one) or, the Hindu art of love* (*see* Penzer, pp. 171–3)
1886. *The perfumed garden of the Cheikh Nefzaoui: a manual of Arabian erotology (XVI century)* (*see* Penzer, pp. 173–7)
1887. *The Behâristân (abode of spring)* (*see* Penzer, p. 177)
1888. *The Gulistân, or rose garden of Sa'di* (*see* Penzer, p. 177)

A number of colleagues have given help with the catalogue in various ways: the late Mrs G. Reiss, and her successor as Leverhulme Librarian, Mrs I. Michaelis; Mrs Beverley Emery and Mrs Marylin Hulme. I am much indebted to them, especially Mrs Reiss and Mrs Michaelis who were responsible for the initial listing. I am also grateful for the help given by Mr Peter Colvin and Miss Z. B. Siddiqi who transliterated and translated material in Arabic and Urdu respectively, and Mr Geoffrey Allen who dealt with the Greek.

July 1978

B. J. Kirkpatrick

1–109 Books, Pamphlets and Translations by, or, edited by Burton

BOOKS AND PAMPHLETS

1 a c d h j*

Goa, and the Blue mountains: or, six months of sick leave. London, Richard Bentley, 1851. xix, 368 pp., front., plates, map. (J. Gerson da Cunha, Bombay, 20 July and 3 Aug. 1877).
First edition, first issue; rebound; *wanting* front. (*see* Penzer, pp. 37–8).

2 c d

Scinde: or, the unhappy valley; second edition. London, Richard Bentley, 1851. 2 vols.
Probably a proof copy; bound in black by Burton (*see* Penzer, pp. 39–40). Extensive annotations; prepared for a 'third edition', *i.e. Sind revisited;* the latter is virtually a new work [*see* 58].

3 a c d g

Sindh, and the races that inhabit the valley of the Indus; with notices of the topography and history of the province. London, Wm. H. Allen, 1851. viii, 422 pp., front. (map), illus.
Extensive annotations; prepared for a second edition which was not published; Burton recorded on the title-page: 2d Edition / revised, partly re-written and containing the / passages suppressed in the First Edition of 1851. (*see* Penzer, p. 40).
Insertions, pp. 172–3: word list in Arabic, Sindhi, and Persian.

4 a c d h j

Falconry in the valley of the Indus. London, John van Voorst, 1852. xv, 107 pp., front., plates. (John van Voorst, 22 Jan. 1877).
Extensive annotations; prepared for a second edition which was not published. IB's autograph on upper free end-paper: Isabel Arundell, Nov. 56. (*see* Penzer, p. 41). The publisher records: Five hundred of the book were printed and 257 remain. I fear the only way of getting rid of them is by making waste paper of them. (*see* inserted letter).
Insertions: publisher's statement – clippings.

5 d

A complete system of bayonet exercise. London, printed and published by William Clowes and Sons, 1853. 36 pp., illus. [*see* 236].

6 a b d h j

Personal narrative of a pilgrimage to El-Medinah and Meccah. London. Longman, Brown, Green, and Longmans, 1855. 3 vols., fronts., illus., plates, maps. (A. Sprenger, Wabern near Berne, 27 Jan. 1871; – Bonola, Venise, [*c.* 1877]).
IB's autograph in Vol. 3, upper free end-paper: Isabel Arundell, Sept 56 (*see* Penzer. pp. 49–50).

7 c d h j

First footsteps in East Africa: or, an exploration of Harar. London, Longman, Brown, Green, and Longmans, 1856. xli, 648 pp., 2 maps, illus. (*lacks* plates). (G. Ferrand, Alger, 28 juin 1884; [Philipp] Paulitscke, Weidlingau bei Wien, le août 1884).
Variant binding, probably a proof or advance copy, bound by Burton. Most of Burton's corrections and notes were incorporated by IB in the *Memorial Edition,* 2 vols., published by Tylston & Edwards, 1894 (*see* Penzer, pp. 60–3).
Insertions: originals of 2 maps – 2 documents in Arabic.

* For an explanation of the lower-case letters at the side of the item number see Preface 7, Indications of Burton's use of a book, p. ix.

8

A coasting voyage from Mombasa to the Pangani river: the expedition to the interior. London, 1858. An offprint of 39 pp., map. Contents: Part 1 [untitled] – Part 2: A visit to Sultan Kimwere, by Burton – Part 3: Progress of the expedition into the interior [of E. Africa], by J. H. Speke. Reprinted from the *Journal of the Royal Geographical Society,* Vol. 28, pp. 188–226.
Unrecorded offprint. [*see* 241].
Another copy. [*see* 238].

9

Zanzibar, and two months in East Africa. [Edinburgh, 1858]. From *Blackwood's Edinburgh magazine,* Feb., March, May 1858, Vol. 83, pp. 200–24, 276–90, 572–89.

10 a d f g h j l

The lake regions of central Africa: a picture of exploration. London, Longman, Green, Longman, and Roberts, 1860. 2 vols. fronts., illus., plates, map.
First edition, second issue (*see* Penzer, pp. 65–6). Also annotated and marked by IB (Vol. 2).
Insertions: 6 holograph pencil sketches (Vol. 1) – 14 holograph notes (unpublished paragraphs and corrections) in Vol. 1, pp. 52, 58, 68, 74, 77, 80, 90, 99 (2 notes), 126, verso of plate facing p. 127, 176, 248, 315 – autograph letters signed: B.B. Walker, Stockwell, [London], S.W., 28 Nov. 1875 – C. I. dal Verme [Civil Engineer], Milan, 8 May 1873 – cuttings.

11 c d g h

The city of the Saints and across the Rocky mountains to California. London, Longman, Green, Longman, and Roberts, 1861. xii, 707 pp., front., illus., plates, map.
Insertions: 10 holograph leaves from original manuscript.

12

Ethnological notes on M. du Chaillu's "Explorations and adventures in equatorial Africa". [London, 1861]. From *Transactions of the Ethnological Society,* N.S. Vol. 1, pp. 316–26.

13 d f g j

Abeokuta and the Camaroons mountains: an exploration. London, Tinsley Brothers, 1863. 2 vols. fronts., plates, map.
Insertions: 2 holograph sketches (Vol. 2) – 4 holograph leaves (scored in pencil) and numbered 37, 38, 75 and 77 appear to be suppressed passages from *Zanzibar,* Vol. 1, pp. 380–1, 464–8 [*see* 40] – manuscript notes, 5 pp., signed: G. G. Bianconi, [Bologna, 2 Aug. 1876] – cuttings.

14

[An account of an] exploration of the Elephant mountain in western equatorial Africa. [London, W. Clowes, printer], 1863. An offprint of 10 pp. Reprinted from the *Journal of the Royal Geographical Society,* Vol. 33, pp. 241–50.
Unrecorded offprint. [*see* 241].

15

A day amongst the Fans [and discussion. London, 1863]. A paper read to the Anthropological Society. Title in contents: A day with the Fans, and on p. 185: A day among the Fans. From the *Anthropological review,* Vol. 1, pp. 43–54, 185–7.

16 d

My wanderings in West Africa: a visit to the renowned cities of Wari and Benin, by an F.R.G.S. [London, 1863]. From *Fraser's magazine,* Feb.–April 1863, Vol. 67, pp. 135–57, 273–89, 407–22. [*see* 218].

17 a c d

The prairie traveler, a hand-book for overland expeditions; with maps, illustrations, and itineraries of the principal routes between the Mississippi and the Pacific, by Randolph B. Marcy . . . edited (with notes) by Richard F. Burton . . . London, Trübner, 1863. xvi, 251 pp., front., illus.
Variant binding in purple cloth boards; lettered in gold on spine within a floral design, and at foot, triple rule in gold at head and foot of spine; triple rule in gold round upper cover; triple rule in blind round lower cover; lacks maps (*see* Penzer, pp. 69–70).

18 a c d f h l

Wanderings in West Africa from Liverpool to Fernando Po, by a F.R.G.S. London, Tinsley Brothers, 1863. 2 vols. fronts., illus., map.
Variant binding in dark purple-brown cloth boards; triple rule in blind round upper and lower covers; lines 5–6 on spine lettered: BY A / F.R.G.S. (*see* Penzer, pp. 71–2).
Extensive annotations in Vol. 1; some marked passages by IB.
Insertions: 5 holograph sketches (Vol. 1, pp. [i–ii], 192; Vol. 2, p. 17).

19 c d h l

A mission to Gelele, King of Dahome, with notices of the so called "Amazons," the grand customs, the yearly customs, the human sacrifices, the present state of the slave trade, and the Negro's place in nature; second edition. London, Tinsley Brothers, 1864. 2 vols. front.

Autograph dedication: To my darling wife – some corrections by Burton – 5 cuttings.

20 d h

The Nile basin. Part I: Showing Tanganyika to be Ptolemy's Western Lake Reservoir; a memoir read before the Royal Geographical Society, November 14, 1864, by Richard F. Burton . . . Part II: Captain Speke's Discovery of the source of the Nile; a review, by James M'Queen . . . London, Tinsley Brothers, 1864. [4], iv, 195 pp., maps. imperfect.

21

Notes on scalping. [London, 1864]. From the *Anthropological review,* Vol. 2, no. 4, pp. 49–52.

22 a

Notes on Waitz's anthropology. [London, 1864]. From the *Anthropological review,* Vol. 2, no. 7, pp. 233–50. [*see* 218].

Reference is made in the *Anthropological review,* 1868, Vol. 6, pp. 462–3 to Burton 'now busily engaged annotating the second volume of Waitz's *Anthropology of primitive peoples,* the English edition of which has been forwarded to him for that purpose by Mr J. Fred. Collingwood . . . The Council of the Anthropological Society is, we believe, ready to order it to be printed immediately the MS. returns to England.' *Introduction to anthropology,* by Theodor Waitz, edited by J. Frederick Collingwood, was published by Longman, Green for the Anthropological Society in 1863 as *Anthropology of primitive peoples,* Vol. 1. Vol. 2 was not published. *See also* 301, 310.

23

On skulls from Annabom, in the West African seas, by R. F. Burton and C. Carter Blake. [London, 1864]. From the *Journal of the Anthropological Society,* Vol. 2, no. 6, pp. ccxxx–ccxxxi. Contents: Letter from Burton to the Secretary of the Anthropological Society, and report by the latter, C. Carter Blake.

24

Reports by Consul Burton of his ascent of the Congo river, in September 1863. [London]. 1864. 9 pp. At head of title: Printed for the use of the Foreign Office. September 6, 1864. Confidential. [*see* 55].

25

A pictorial pilgrimage to Mecca and Medina (including some of the more remarkable incidents in the life of Mohammed, the Arab lawgiver). London, printed for the author, by William Clowes & Sons, 1865. 58 pp., front. (port.). At head of title: The guide-book.

d

Another copy, probably page proofs; wanting pp. 5–16; extensive annotations. [*see* 220].

26 c d g h i

Stone talk . . . being some of the marvellous sayings of a petral portion of Fleet Street, London, to one Doctor Polyglott, Ph.D., by Frank Baker, D.O.N. London, Robert Hardwicke, 1865. [iii], 121 pp.

Title-page imperfect; lacks errata slip (*see* Penzer, p. 77). 200 copies printed (*see* inserted statement). Extensive annotations and corrections.

Insertions: 4 holograph leaves of blank verse (1 loose, 3 at end) – autograph letter signed: Robt Hardwicke, [London], 29 April 1865 – publisher's statement – list of journals and clubs sent copies – 4 cuttings (1 loose).

c k l

Another copy (IB's); autograph dedication: From a Friend; errata slip present.

27 a c d h

Wit and wisdom from West Africa: or, a book of proverbial philosophy, idioms, enigmas, and laconisms, compiled by Richard F. Burton. London, Tinsley Brothers, 1865. xxxi, 455 pp.

Variant binding in mauve cloth boards; no 'rough sheet of notes' as recorded by Penzer (*see* p. 76).

28 d

From London to Rio de Janeiro: letters to a friend. [London, 1865–6]. From *Fraser's magazine for town and country,* Vols. 72–3, nos. 430, 433, 436, pp. 492–503, 78–92,

496–510 respectively. Contents: Letter[s] I: To Lisbon – II: Lisbon, A.M. – III: Lisbon, P.M. – IV–V: Cape Verde islands – VI: 'The Reef' (Arrival At). Extensive annotations. Letters V–VI unrecorded.
Another copy (Letters I–IV only). [*see* 220].

29
On a kjökkenmödding at Santos, Brazil. [London, 1866]. From the *Journal of the Anthropological Society,* Vol. 4, pp. cxciii–cxciv. Letter to the Anthropological Society, dated Santos, Brazil, 11 Dec. 1865.

30
The extinction of slavery in Brazil, from a practical point of view. [London, 1868]. From the *Anthropological review,* Vol. 6, pp. 56–7. Letter to the Anthropological Society, dated Rio de Janeiro, 1 June 1867. Pp. 57–63 translation by Richard Austin of A. M. Perdigao Malheiro's communication in the *Jornal do commercio,* Rio de Janeiro, 17 April 1867, on which Burton comments in his letter.

31 c d h
The highlands of Brazil. London, Tinsley Brothers, 1869. 2 vols, fronts., illus., map. Second title-page: Explorations of the highlands of Brazil, with a full account of the gold and diamond mines; also canoeing down 1500 miles of the great river São Francisco, from Sabará to the sea.
First edition, first issue (*see* Penzer, p. 79).
Autograph signature: Isabel Burton Jany 1st 69 – 9 cuttings.

32 d h
Letters from the battle-fields of Paraguay. London, Tinsley Brothers, 1870. xix, 491 pp., front., map. Second title-page.
First edition, first issue; no 'long letter from Burton' as recorded in Penzer (*see* Penzer, pp. 84–5).
Insertions: cuttings (incl. 9 loose in box).

33 c g h j
Vikram and the vampire: or tales of Hindu devilry, adapted by Richard F. Burton, with thirty-three illustrations by Ernest Griset. London, Longmans, Green, 1870. xxiv, 319 pp., front., illus., plates.
First edition, first issue (*see* Penzer, p. 82). 3000 copies printed (*see* inserted statement). *Insertions:* 9 holograph leaves (8 loose) – autograph letter signed: Longmans, Green, [London], 20 June 1884 – publisher's statement – 9 cuttings (4 loose). *See also The Kathá sarit ságara,* translated by C. H. Tawney, and *The Buttris Singhasun,* translated by Cheedam Chunder Das [1042, 1079].

34
Copy of a letter to Her Majesty's Secretary of State for Foreign Affairs, signed Richard F. Burton, and dated 14 Montagu Place, Montagu Square W. 16th October 1871. 8 pp. and 29 pp. comprising copies of 34 letters of regret at Burton's recall.
On his recall from the Consulship of Damascas (*see* Penzer, p. 215). [*see* 219].

35
Proverbia communia syriaca. [London, 1871]. An offprint of 29 pp. Reprinted from the *Journal of the Royal Asiatic Society,* N.S., Vol. 5, pp. 338–66; reprinted in *Unexplored Syria,* Vol. 1 as Appendix II, pp. 263–94.
Unrecorded offprint. [*see* 220].

36
The revival of Christianity in Syria: its miracles and martyrdoms, related by P. London Edward Stanford, [1871]. 24 pp.
Published by IB under her own name in her *The inner life of Syria, Palestine, and the Holy land,* 1875, Vol. 1, pp. 180–200, and in her *The Life,* Vol. 1, pp. 548–67; she records the circumstances of its first appearance on p. 548 of the latter. There are additions and deletions in three printings. Not in Penzer.
Another copy. [*see* 219].

37
[An altar-stone from the site of the ancient Canatha, in Jebel Duruz Hauran, and A thurible of bronze, found in the country between Palmyra and Damascus, exhibited by Captain R. F. Burton. London, 1872]. An offprint of 3 pp., 1 plate. Reprinted from the *Proceedings of the Society of Antiquaries of London,* Series 2, Vol. 5 pp. 289–91.
Unrecorded offprint. [*see* 220]. Another copy. [*see* 39].

38
A ride to the Holy land. [London, 1872]. From *Cassell's magazine,* N.S., Vol. 5 pp. 434–7. [*see* 221].

39 a c d h j l

Unexplored Syria: visits to the Libanus, the Tulúl el Safá, the Anti-Libanus, the northern Libanus, and the 'Aláh, by Richard F. Burton and Charles F. Tyrwhitt Drake. London, Tinsley Brothers, 1872. 2 vols. fronts., illus., plates, map.
First edition, first issue (*see* Penzer, p. 86); also carries autograph signature: Isabel Burton; very scattered annotations and corrections in Vol. 1, pp. 198, 229, 360, plates 1 and 4 facing p. 360, and Vol. 2, pp. 5, 44, 77, 137; marked passages probably by IB.
Insertions: sketch signed: E. J. Davis, Alexandria (Vol. 1, plate facing p. 360) – autograph letter signed: Ch. Clermont Ganneau, Paris, 13 May 1872 – pamphlet [*see* 37] – cuttings.

40 a c e h j

Zanzibar: city, island, and coast. London, Tinsley Brothers, 1872. 2 vols. fronts., plates, maps.
Rebound; a few corrections.
Insertions: autograph letter signed: George Percy Badger, [London], 21 Feb. 1872 – 3 cuttings – review of book. *See also* 13 for suppressed passages.

41

En route to Hebron. [London, 1873]. From *Cassell's magazine,* 1873, N.S., Vol. 6, pp. 69–72. [*see* 221].

42

The primordial inhabitants of Minas Geraes, and the occupations of the present inhabitants. [London, 1873]. An offprint of [17] pp. Reprinted from the *Journal of the Anthropological Institute,* Vol. 2 (1873) for 1872, pp. 407–23.
Unrecorded offprint. [*see* 220].

43

Supplementary papers to the Mwátá Cazembe (Journal of Dr de Lacerda) by the translator Richard F. Burton. [Trieste, 1873]. xliii pp.
These papers, considered controversial by the Royal Geographical Society, were omitted from their edition of *The lands of Cazembe,* 1873. They comprise 'Preface by the Translator' – Appendix I: Notes on *How I found Livingstone in central Africa . . .* by Henry M. Stanley – and Appendix II: Being a rejoinder to the 'Memoir on the Lake regions of East Africa reviewed', in reply to Captain Burton's letter in *The Athenaeum,* No. 1899, by W. D. Cooley. Burton notes on p. [iii] 'I have struck off, especially for the use of my friends, a few copies of the rejected matter.' [*see* 217].
Another copy; autograph signature: Isabel Burton. [*see* 220].

44 a d h

The captivity of Hans Stade of Hesse, in A.D. 1547–1555, among the wild tribes of eastern Brazil; translated by Albert Tootal . . . and annotated by Richard F. Burton. London, printed for the Hakluyt Society, 1874. xcvi, 169 pp., bibliogr. (Hakluyt Society [Publications], Vol. 51).
Annotations on upper paste-down end-paper, pp. v, [14].
Insertions: list of presentation copies – 3 cuttings.

45*a* l

Notes on the castellieri or prehistoric ruins of the Istrian peninsula. [London, 1874]. Paper read to the London Anthropological Society, of which Burton was Vice-President. From *Anthropologia,* Vol. 1, pp. 376–404 text; pp. 404–8 discussion. P. 375 'A Word to the Reader', letter to the London Anthropological Society dated Trieste, Oct. 1874; pp. 408–15 Appendix: I castellerieri dell'Istria, by S. – Burton's reply to S.

45*b*

— [London, 1874]. An offprint of 40 pp., 3 plates. Reprinted from *Anthropologia,* Vol. 1, pp. 375–415. [*see* 220].

46

Notes on Rome. [London, 1874]. From *Macmillan's magazine,* Nov.–Dec. 1874, pp. 57–63, 126–34.

47

Geographical notes on the Province of Minas Geraes, by Henrique Gerber; translated and communicated by Captain Richard F. Burton. London, 1875. An offprint of 40 pp., tables. Reprinted from the *Journal of the Royal Geographical Society,* 1874, pp. 262–300. [*see* 220]. *See also* 1821.
Unrecorded offprint.
Another copy; and 2 sets of proofs dated 6 and 7 Jan. 1875. [*see* 217].

48

The port of Trieste, ancient and modern. London, 1875. From the *Journal of the Society of Arts,* 29 Oct. and 5 Nov. 1875, Vol. 23, pp. 979–86, 996–1006. [*see* 221].

49

[Speech by Burton at the Annual Meeting of the Palestine Exploration Fund, on the Trans-Jordan region. London, 1875]. [1] leaf. Proof of speech published in the *Palestine Exploration Fund quarterly statement,* July 1875, pp. 120–3. [*see* 220].

50 a c f g h j

Ultima Thule: or, a summer in Iceland. London, William P. Nimmo, 1875. 2 vols. fronts., illus., plates, map.

Insertions: holograph notes, 6 leaves – original sketches for text figures in Vol. 1, pp. 303–6 (2 leaves), for plate facing Vol. 2, p. 279 (2 leaves) – 13 holograph letters signed: W. P. Nimmo, Edinburgh, 11 July 1876 – A. Sprenger, Wabern, 1 Nov. 1879 and Kissingen, 14 Aug. 1880 – F. W. Simms, Cairo, 7 Dec. 1887 – Jón A. Hjaltalín, Edinburgh, 10 Feb. 1877 – R. M. Smith, Edinburgh, 7 Oct. 1876 – Chatto & Windus, publishers, London, 11 May 1877 – Samuel Tinsley, publisher, London, 7 and 14 June 1877 – Edward Stanford, London, 22 June 1877 – Longmans & Co., London, 9 and 12 July 1877 – G. Fitz Roy Cole, [London], 17 July 1877 (the last 7 concern the sequel *Iceland revisited* which was not published) – 2 statements from C. Kegan Paul, 31 Dec. 1880 – cuttings.

51 a c d g h j

Etruscan Bologna: a study. London, Smith, Elder, 1876. xii, 275 pp., front., illus.

A black decorative border runs two thirds across the upper cover, within a double rule, near head and foot; bevelled edges (*see* Penzer, p. 92). 1500 copies printed (*see* inserted letter from Smith, Elder, 27 Oct. 1875).

Insertions: originals of 21 illus., annotated by Burton – 13 holograph leaves – 3 manuscript leaves, annotated by Burton – telegram, Bologna, [7 Nov.] 1875 – map of Bologna – autograph letters signed: G. Ge. or G. Giuse. Bianconi, Bologna, 30 April and 25 Nov. 1875, 10 and 29 Dec. 1876, 28 Feb. and 2 May 1877; C. Carter Blake, undated; Placido Brandoli, Rome, 1875 and Modena, 21 April and 28 Sept. 1877; Luigi Calori, Bologna, 9 Nov. 1876; G. Capellini, Museo di Geologia e Paleontologia, Bologna, 16 Dec. 1876; Hyde Clarke, [London], 20 Jan. 1877; Saml. Ferguson, Public Record Office of Ireland, Dublin, 18 Feb. 1877; G. Gozzadini, Bologna, 6 and 22 Feb. 1875, 28 Nov. and 23 Dec. 1876; Camillo Munzi, Bologna, 29 Dec. 1876; Smith, Elder & Co., London, 27 Oct. 1875, 1 Nov. 1876, 3 Jan., 6 Feb. and 29 May 1877; E.O.E. Stillmans, Ventnor, 1 Feb. [187-]; A. Zannay, Bologna, 30 Jan. 1877; 1 unsigned, Trieste, Sept. 1875 – cuttings.

52

Fire-arms and projectiles. London, George E. Eyre & William Spottiswoode, printers; published at the Great Seal Patent Office, 1876. 2 pp. At head of title: A.D. 1875, 1st September. No. 3069. Title on wrapper: Specification of Captain Richard Burton.

53

The long wall of Salona and the ruined cities of Pharia and Gelsa di Lesina. [London, 1876]. An offprint of [49] pp., illus., 2 plates. Reprinted from the *Journal of the Anthropological Institute,* Vol. 5 (1876) for 1875, pp. 252–99.

Unrecorded offprint. [*see* 220].

Two other copies. [*see* 217].

54 a c d f g h j

A new system of sword exercise for infantry. London, William Clowes, 1876. 59 pp., illus.

Variant binding lacking design in gold on upper cover (*see* Penzer, p. 93). Prepared for a second edition; holograph note on title-page: IId Edition (with additions and improvements). 500 copies printed (*see* 107, Box 2, Item 21, letter from William Clowes, 4 Aug. 1876).

Insertions: 6 original illus. (2 by Burton) – 9 pp. (holograph) for insertion in pp. 18–19 (6 pp. and 3 pp.) – 2 leaves (annotated) – autograph letter signed: A[rchibald] Maclaren, Oxford, [1875] – 2 pamphlets. [*see* 1432, 1447].

55 a c d f g h

Two trips to gorilla land and the cataracts of the Congo. London, Sampson Low Marston, Low, Searle, 1876. 2 vols. fronts., illus., plates, map.

Variant binding in brown cloth (*see* Penzer, p. 94).

Insertions: 112 holograph leaves (part of original and some unpublished material) – 1 holograph sketch – Appendix IV (in part) – extract from *O Muata Cazembe,* 3 leaves – pamphlet [*see* 24] – map – photograph (scenic) – publisher's agreement – cuttings.

56*a*

The volcanic eruptions of Iceland in 1874 and 1875 . . . (with two maps of Iceland)

[Edinburgh, 1876]. An offprint of [15] pp. Reprinted from the *Proceedings of the Royal Society of Edinburgh,* Session 1875/6, Vol. 8, pp. 44–58.

56*b*

The volcanic eruptions of Iceland in 1874 and 1875. [Edinburgh, 1876]. An offprint of 14 pp., illus. Reprinted from the *Proceedings of the Royal Society of Edinburgh,* Session 1875/6, Vol. 8.

Another copy, preceded by 'Notes on human remains brought from Iceland by Captain Burton', by C. Carter Blake. [*see* 220].

57

Scoperte antropologiche in Ossero. [Trieste, 1877]. An offprint of 6 pp., illus. Reprinted from *Archeografo triestino,* Vol. 5, no. 2, pp. 129–34.

Another copy. [*see* 220].

58 c d h j l

Sind revisited: with notices of the Anglo-Indian army, railroads, past, present and future, etc. . . . London, Richard Bentley and Son, 1877. 2 vols.

Marked passages probably by IB.

Insertions: autograph postcard and letters from the publisher dated 28 Dec. 1876, 11, 27 Jan., and 5 Feb. 1877.

59

Flint flakes from Egypt. [London, 1878]. An offprint of 2 pp., illus. Reprinted from the *Journal of the Anthropological Institute,* Vol. 7, pp. 323–4. Unrecorded offprint.

Another copy. [*see* 220].

60 a c d f g j

The gold-mines of Midian, and the ruined Midianite cities: a fortnight's tour in north-western Arabia. London, C. Kegan Paul, 1878. xvi, 398 pp., illus., map, *and* facsimiles (7 copies) of illus. on pp. 209–241.

1000 copies printed (*see* statement in 50).

Insertions: holograph sketches (verso of map, also heavily annotated) – 8 holograph leaves (Preface to The land of Midian (revisited); *see* 64) – autograph letters signed: E. Beke, Harrogate, [1878]; C. G. Cleminshard, Lessney Heath, North Kent, 12, 16 Oct. 1880, 16 June 1881; Robert Fleming, Cairo, 30 Jan. 1880, including reports on Egyptian petroleum and oil – W. A. Ross, [London], 25 June 1878; A. Sprenger, Wabern near Bern, 5 Feb. and 16 March 1879 – pamphlet [*see* 1294] – cuttings.

61

Midian and the Midianites. London, 1878. From the *Journal of the Society of Arts,* 29 Nov. 1878, Vol. 27, pp. 16–27. [*see* 221].

62

More castellieri, by Richard F. Burton and Messieurs Antonio Scampicchio . . . and Antonio Covaz. [London, 1878]. An offprint of 23 pp. Reprinted from the *Journal of the Anthropological Institute,* Vol. 7 (1878) for 1877, pp. 341–63.

Unrecorded offprint. [*see* 220].

63

Episode of Dona Ignez de Castro (The Lusiads of Camoens). Canto III, stanzas 118–135; printed for private circulation. London, Harrison and Sons, 1879. 7 pp.

IB refers to this pamphlet as a sample (*see* her *The Life,* Vol. 2, p. 181). Each stanza is a variant of either the translation published in *The Lusiads* [*see* 75*a*] or of capitalisation and punctuation.

Another copy. [*see* 220].

64 a c d h j

The land of Midian (revisited). London, C. Kegan Paul, 1879. 2 vols. fronts., illus., plates, map.

1000 copies printed (*see* statement in 50).

Insertions: autograph postcard and letters signed: G.P.B. and George Percy Badger dated [London], 23 Jan. and 14 June [187–]; C. Carter Blake, [London], 14 Oct. 1878; A. Günther, British Museum, 25 Oct. 1878; W. E. Haynes of Greenfield & Co., Alexandria Harbour Contract, 30 June 1877; Curie Morisol, Alexandria, 25 Oct. 1878; James Scott, Yarek, Victoria, 4 and 18 July 1878 – pamphlet [*see* 1377] – cuttings.

65

The Ogham-runes and el-Mushajjar: a study. [London, 1879]. An offprint of 46 pp., illus., 4 plates. Reprinted from the *Transactions of the Royal Society of Literature,* Vol. 12.

See also Penzer, p. 235, and *The Life,* Vol. 2, p. 134, note.

Another copy, wanting plates. [*see* 220].

66

Remains of buildings in Midian. [London, 1879]. An offprint of [22] pp., 2 plates. Reprinted from the *Transactions of the Royal Institute of British Architects*, Session 1878/9, No. 3, pp. 61–84. Contents: pp. 61–80 text – pp. 81–2 discussion – pp. 83–4 Gold in Midian, which also appeared in *The Athenaeum*. Unrecorded offprint.
Three other copies.

67

Stones and bones from Egypt and Midian. [London, 1879]. An offprint of 30 pp., 2 plates. Reprinted from the *Journal of the Anthropological Institute*, Vol. 8, pp. 290–319. Unrecorded offprint. [*see* 220].

68

Veritable and singular account of an apparition, and the saving of a soul, in Castle Weixelstein, in Krain. [London, 1879]. From *The Spiritualist*, 19 Sept. 1879, No. 369, pp. 138–40. Reprinted in *The Life*, Vol. 2, pp. 170–3, with a preliminary account on pp. 168–9. [*see* 219, 120].

69 d

Correspondence with His Excellency Riaz Pasha upon the mines of Midian; printed for private circulation only. Alexandria, Egypt, The Alexandria Stationers' & Booksellers' Company, 1880..14 pp. Another copy. [*see* 220].

70 d h

The ethnology of modern Midian. [London, 1880]. 82 pp. Page proofs; published in the *Transactions of the Royal Society of Literature*, Series 2, Vol. 12, pp. 249–330.

71

Giovanni Battista Belzoni. [London, 1880]. An offprint of 15 pp. Reprinted from the *Cornhill magazine*, July 1880, Vol. 42, pp. 36–50. Unrecorded offprint. [*see* 220].

72

Itineraries of the second expedition into Midian. London, [1880]. An offprint of 150 pp., front. (map). Reprinted from the *Journal of the Royal Geographical Society*, for the Session 1878/9, Vol. 49. Title in the latter; Itineraries of the Second Khedivial Expedition: Memoir explaining the new map of Midian made by the Egyptian Staff Officers.

73

Os Lusiadas (The Lusiads) rendered into English, by Richard Francis Burton; edited by his wife Isabel Burton. London, Tinsley Brothers, 1880. xviii pp. The text comprises *Os Lusiadas*, Vol. 1, pp. [i]–xix, published by Bernard Quaritch (*see* 75). Probably issued as a sample (*see* note to 74). [*see* 219].

74 d

The Lusiads. [London, 1880]. 7 pp. The text comprises Argument and Canto 1, Stanzas 1–8 from the *Os Lusiadas*, Vol. 1, pp. [3]–7; each stanza, however, is a variant of the published version (*see* 75*a*). IB refers to this pamphlet as a sample (*see* her *The Life*, Vol. 2, p. 181) [*see* 75*b*].
Two other copies, both variant issues. [*see* 219–20].

75*a* c l

Os Lusiadas (The Lusiads): Englished by Richard Francis Burton; (edited by his wife, Isabel Burton). London, Bernard Quaritch, 1880. 2 vols.
Variant binding as second issue but lacks gilt top edges (*see* Penzer, p. 104). Annotations and marked passages by Gerald Massey.

75*b* c d h j

— Another copy bound by Burton in one volume in maroon cloth boards, lettered in gold on upper cover; imprint on title-page: LONDON: / TINSLEY BROTHERS, / CATHERINE STREET, STRAND, W.C. / 1880. / ALL RIGHTS RESERVED.
Heavily annotated, with corrections and additions for a revised edition.
Insertions: autograph letters signed: J. Y. Johnson, Funchal, Madeira, 12 May 1884 Edgar Prestage, Balliol College, Oxford, 5 Feb. 1889, and Bowden, Chesire, 10 Aug. 1889 – The Lusiads, 7 pp. – cuttings.

75*c* a c l

— Another copy of Vol. 1 probably bound by Burton in dark brown paper boards dark brown cloth spine and corners; lettered in gold on spine with blind and gold rules. Bernard Quaritch imprint. Holograph note by IB: Gerald Masseys corrections.

Bernard Quaritch's title-page is a cancel (*see* 75*a*); it would, therefore, appear that this work was to have been published by Tinsley Brothers. A few of Burton's corrections in the Tinsley Brothers's edition (75*b*) are listed in the errata slip tipped-in the Quaritch issue, Vol. 1, [1].

76*a* **d**

Report on two expeditions to Midian. Alexandria, Egypt, The Alexandria Stationers' & Booksellers' Company Limited, 1880. 6 pp. P. 6 at foot: Alexandria, January 1, 1880.

76*b*

— Alexandria, Egypt, 1880. 7 pp. P. 7 at foot: Alexandria, January 1, 1880. Text slightly revised. [*see* 220].

76*c*

— London, 1880. From the *Journal of the Society of Arts,* 31 Dec. 1880, Vol. 29, No. 1467, pp. 98–9. Text further revised.
Another copy.

77

Report upon the minerals of Midian. Alexandria, The Alexandria Stationers' & Booksellers Coy Lt., 1880. 16 pp. P. 16 at foot: Shepheard's, Cairo, January 30, 1880. *See also* Preface to *The land of Midian (revisited)* (64). [*see* 220].

78

A visit to Lissa and Pelagosa. London, 1880. An offprint of 40 pp. Reprinted from the *Journal of the Royal Geographical Society,* 1879, Vol. 49, pp. 151–90.

79*a* **c j**

Camoens: his life and his Lusiads: a commentary, by Richard F. Burton. London, Bernard Quaritch, 1881. 2 vols. Lettered on spine: The Lusiads. Vol. III [Vol. IV]. First issue (*see* Penzer, p. 104).
Insertions: autograph letters signed: Edgar Prestage, Oxford, 24 April 1888, and Bowden, Cheshire, 8 Jan. 1889.

79*b* **c d g h j**

— Proof copy, bound in one volume, probably by Burton, in dark brown paper boards, cloth spine and corners; lettered in gold on spine with blind and gold rules. Heavily annotated with corrections and additions for a revised edition.
Insertions: 13 holograph leaves bound inside upper cover, and 10 leaves (5 holograph) bound inside lower cover – autograph letter and notes signed: H. Yule, [London], 1 Nov. 1881 and 15 July 1885 – list of editions of *The Lusiads* – cuttings.

80 c

A glance at the "Passion-play". London, W. H. Harrison, 1881. 168 pp. front., plans, music score.

81

How to deal with the slave scandal in Egypt; reprinted from the *Manchester examiner and times,* March 21st, 23rd, and 24th, 1881. [Trieste, 1881]. 32 pp. Signed A.E.I. [*i.e.* IB]. This work is by Burton, and appears under his own name in *The Life,* Vol. 2, pp. 195–210. [*see* 219].

82

The partition of Turkey; reprinted from the *Manchester examiner and times,* January 3rd and 4th, 1881. [Trieste, 1881]. 23 pp. Signed A.E.I. [*i.e.* IB]. This work is by Burton, and appears under his own name in *The Life,* Vol. 2, pp. 552–9.
Two other copies [for the 3rd copy *see* 219].

83 d

The thermae of Monfalcone (aqua Dei et vitae). London, Horace Cox, "The Field" Office, 1881. An offprint of 22 pp., plan. Reprinted from *The Field,* 12 Nov., 17 and 24 Dec. 1881.

84

Three sonnets from Camoens. [London, 1881]. [1] leaf. Proof of contribution to *The Athenaeum,* 26 Feb. 1881, p. 299. Variant translations of *The Lyricks,* Part 1 (1884), Nos. i, xix, c, pp. 1, 37, 91. [*see* 219].

85 a d

The gold fields of West Africa, by Captain Cameron, R.N. – Gold on the Gold Coast, by Captain Richard F. Burton. London, printed by W. Trounce, 1882. An offprint of 19 pp. Reprinted from the *Journal of the Society of Arts,* Vol. 30, pp. 777–94. Unrecorded offprint. [*see* 219].

86

Lord Beaconsfield: a sketch. [London, ?1882]. 12 pp. IB records in her *The Life,* Vol. 2, p. 212 'Richard wrote a "sketch", which made twelve pages of print, which will appear in "Labours and Wisdom."' [*see* 219].

87

On stone implements from the Gold Coast, West Africa, by Captain R. F. Burton . . . and Commander V. L. Cameron. London, 1883. An offprint of 6 pp. illus. Reprinted from the *Journal of the Anthropological Institute,* Vol. 12, pp. 449–54. Unrecorded offprint.

88 a c d g h j
To the Gold Coast for gold: a personal narrative, by Richard F. Burton and Verney Lovett Cameron. London, Chatto & Windus, 1883. 2 vols. fronts., illus., maps; *wanting* Vol. 2, front.
Insertions: 1 holograph leaf – autograph letter signed: H. Boudent, Saumur, 26 June 1883, on translating the book – cuttings. *See also* 106 for manuscript.

89*a*
The book of the sword. London, Chatto and Windus, 1884. xxxix, 299 pp. illus. Variant binding lacking publisher's device on lower cover (*see* Penzer, p. 108). *See also* 107 for material intended for two further volumes.

89*b* d j l
— Proof copy dated 1883.
Insertions: autograph letter and post-card signed: A. Pitt-Rivers (formerly Lane-Fox), [London], 13 March 1883; [?], post marked London, 19 March 1883 – copy for most of the illustrations tipped-in.

89*c* d g l
— Proof copy dated 1883 comprising several imperfect sets; stamped on title: 1 Nov. 83. Corrections by printer's reader except pp. 199–200, 274–7, 279–80; revise at end dated 22 Nov. 1883.
Insertion: 1 holograph leaf tipped-in p. 175 (first revise).

89*d* d g h j l
— Proof copy dated 1884; bound by Burton, and lettered in gold on upper cover: SWORD. Corrections by Burton and printer's reader.
Insertions: inside upper cover (11 inserts): 4 holograph leaves; 4 autograph letters signed: St. Clair Baddeley, [London], 11, 24 Sept. and 18 Oct. 1884; W. Wareing Faulder, Manchester, 4 June 1884; copy of notes on Sir William Wallace's sword 5 cuttings – holograph leaf tipped-in half-title – pp. 106–7 from *African wanderings*, by F. Were, pp. 173–4, annotated.

90 a c g h j
Camoens, The Lyricks . . . Englished by Richard F. Burton. London, Bernard Quaritch, 1884. 2 vols.
500 copies printed (*see* inserted account).
Insertions: 1 holograph leaf – autograph letters signed: St Clair Baddeley, [London], 18 March 1885 – O. Crawfurd, Oporto, 16 Jan. 1885 – Lallah [*i.e.* Alice Bird] London, 4 Dec. 1884 – A. H. Sayce, Oxford, 8 Aug. 1885 – Archibald [?], Corfu, Xmas 1882 – A. C. Swinburne, 13 April 1881 (to IB) and 27 Nov. 1884 – Quaritch's list of complimentary copies – Wyman & Sons account – cuttings.

91*a* a c d g h j l
The book of the thousand nights and a night, with introduction, explanatory notes on the manners and customs of Moslem men and a terminal essay upon the history of The nights, by Richard F. Burton. Benares, printed by the Kamashastra Society for private subscribers only, 1885–[7]. 10 vols.
Vols. 1 and 10 heavily annotated. Vol. 1, p. [i] annotated: 'Richard Francis Burton with the congratulations of the one most interested in its success – his best friend Sept 10th 1885 London'. Vol. 10 is a proof copy; p. [1] autograph signature: Isabel Burtons copy.
Insertions: Vol. 1: 1 holograph leaf – Prospectus for the work – cuttings. *Vol. 2:* 3 insertions – cuttings. *Vol. 3:* cuttings and advertisements for *The Lusiads* and *The Kasidah*. *Vol. 4:* cuttings. *Vol. 5:* 11 illus. (in envelope) and copy for illus. on p. [i] of each volume. *Vol. 7:* autograph letter signed: A Reader, Paris, 28 Dec. 1886 – holograph note by IB(?). *Vol. 8:* 2 holograph leaves (tipped-in p. [v]) – 2 holograph leaves (tipped-in p. [vii]) – 2 holograph leaves (tipped-in p. 33). *Vol. 9:* cuttings. *Vol. 10:* autograph letter (with portrait) signed: Alexander J. Cotheal, New York, 25 Aug. 1887 and 10 Feb. 1888 – errata-slip tipped-in p. 302 – cuttings.
See also 677–703, especially 677–9, 693–5 for translations used by Burton which he notes in his 'the Translator's foreword', Vol. 1, pp. xix–xx and IB in her *The Life* Vol. 2, p. 285.

91*b* c h
Lady Burton's edition of her husband's Arabian nights translated literally from the Arabic; prepared for household reading by Justin Huntly McCarthy. London Waterlow, 1886–7. 6 vols. fronts. (Vols. 1–2, ports.).

92
[The book of the thousand nights and a night – The translator's foreword. Benares 1885]. vii–xxiv pp. From *The book of the thousand nights and a night,* Vol. 1 pp. [vii]–xxiv. Possibly a preprint.

93

Iraçéma, the honey-lips: a legend of Brazil, by J. de Alencar; translated, with the author's permission, by Isabel Burton. London, Bickers & Son, 1886. vii, 101 pp.
bound following the above:
Manuel de Moraes: a chronicle of the seventeenth century, by J. M. Pereira da Silva; translated by Richard F. and Isabel Burton. Bickers & Son, 1886. viii, 138 pp.
Title on wrapper: Iraçéma or honey-lips & Manuel de Moraes, The convert; translated from the Brazilian by Richard and Isabel Burton.

94*a* **c d g h j l**

Supplemental nights to The book of the thousand nights and a night, with notes anthropological and explanatory by Richard F. Burton. Benares, printed by the Kamashastra Society for private subscribers only, 1886–8. 6 vols.
Insertions: Vol. 1: autograph letters signed: Lallah [Alice Bird], London, 23 Sept. [188–]; H. Graetz, Breslau, 28 Sept. 1884; Dr W. Pertsch, Gotha, 17 July 1886 – Prospectus (2) for "Supplemental nights" – autograph letter to Miss [Alice] Bird, signed: J.H.G. Stone, [Cambridge] – manuscript leaf in Arabic, a quotation from a note on Aden, by al-Qazvīnī (Zakariyā ibn Muhammad ibn Mahmud) – 3 unsigned manuscript leaves – cuttings. *Vol. 3:* autograph letters signed: J. F. Blumhardt, Cambridge, 7 Feb. 1888; W. F. Kirby, London, 18 Dec. 1887 – manuscript leaf by W. F. Kirby (?) p. 119; H. Zotenberg, Paris, 7 Feb. 1888. *Vol. 4:* 2 holograph leaves – 1 unsigned manuscript leaf – autograph letters signed: J. Bellamy, Oxford, 6 Nov. 1886; Edward B. Nicholson, Oxford, 1 Nov. 1886; Salisbury, Hatfield, 1 Dec. 1886 (*see* pp. 357–61, and 2538) – Prospectus for *Supplemental nights* – cuttings. *Vol. 6:* autograph letters signed: Alexander J. Cotheal, New York, 7 and 25 Oct. 1888; Wm. H. Pars [Bookseller], New York, 16 and 19 Oct. 1888 – Prospectus for S. L. Wood's 100 pictures to illustrate Burton's 'Translation of the Arabian Nights', announced for publication by Pickering & Chatto – IB's holograph note on expenses – 2 manuscript leaves by O. V. Houdas(?) – errata slip (2) – cuttings.
Pickering & Chatto did not publish Burton's *Arabian nights,* illustrated by S. L. Wood; they did, however, publish *The Arabian nights' entertainments,* illustrated by S. L. Wood, from the text of Dr Jonathan Scott, 4 vols., in 1892.

94*b*

The biography of The book and its reviewers reviewed. [Benares, 1888]. 385–500 pp.
From *Supplemental nights,* Vol. 6.

95 **a e h**

The Gulistân: or rose garden of Sa'di, faithfully translated into English. Benares, printed by the Kama Shastra Society for private subscribers only, 1888. viii, 282 pp.
Second issue (*see* Penzer, p. 177). Penzer records that *The Gulistân* was translated by Edward Rehatsek (*see* Penzer, p. 162). *See also* 2550.

96 **a h**

Three months at Abbazia. [Vienna], 1888. An offprint of 20 pp. Signed Richard & Isabel Burton, dated London, 4 Aug. 1888. Reprinted from the *Vienna weekly news,* 24, 31 July, and 4 Aug. 1888. *See* inserted clippings from the *Vienna weekly news,* 30 Oct. 1888 for comment on the article.
Two sets of incomplete galley proofs.
Another copy.

97

Il Pentamerone: or, the Tale of tales; being a translation by the late Sir Richard Burton, K.C.M.G., of Il Pentamerone . . . of Giovanni Battista Basile . . . London, Henry and Co., 1893. 2 vols.
'Large paper issue' limited to 165 copies of which Vol. 1 is No. 153, and Vol. 2 is No. 156 (*see* Penzer, p. 155). *See also* 738–44.

98

The Carmina of Caius Valerius Catullus; now first completely Englished into verse and prose, the metrical part by Capt. Sir Richard F. Burton . . . and the prose portion, introduction, and notes explanatory and illustrative by Leonard C. Smithers. London, printed for the translators, in one volume, for private subscribers only, 1894. xxiii, 313 pp., front.
Edition limited to 1,000 copies of which this is No. 235 (*see* Penzer, p. 157). *See also* 806.

99

The sentiment of the sword: a country-house dialogue, by the late Captain Sir Richard F. Burton . . . edited, with notes, by A. Forbes Sieveking . . . and a preface by

Theodore A. Cook. London, Horace Cox, "Field" Office, 1911. xv, 151 pp., front. (port.). Reprinted from *The Field,* 7 May–3 Dec. 1910 (*see* Penzer, p. 247).

MAPS

100
Africa. West coast. Brass & St. Nicholas rivers sketch survey, by Captn. Burton, Bomy. Army & Lieut. [W.D.M.] Dolben, R.N. 1861. London, Admiralty, 1862. 1 sheet.
Another copy.

101
Africa. West coast. River Ogun or Abbeokúta sketch survey, by Comr. [Norman B.] Beddingfield, R.N., & Captn. Burton, Bomy. Army, 1861. London, Admiralty, 1862. 1 sheet.
Another copy.

102 d l
Carte d'Akabah à Maouelah indiquant la situation des mines et des puits et cours d'eau de cette région d'après la reconnaissance faite par Amer Rouchdi . . . et Yusuf Tewfick sous le commandement du Capitaine Burton. [*c.* 1875]. 1 sheet (rolled; holograph). At side of title: North Midian. [*see* map cabinet].

TRANSLATION

103 a c d
Voyages du Capitaine Burton à la Mecque aux Grands lacs d'Afrique et chez les Mormons; abrégés par J. Belin-de Launay d'après le texte original et les traductions de Mme H. Loreau. Paris, Hachette, 1872. xvi, 336 pp., plates, maps.

MANUSCRIPTS

104*a*
Akhlák i Hindi, or, a translation of the Hindústání version of Pilpay's fables, by R. F. Burton, Lt. 18th Regt. Bombay, N.I., with explanatory notes, and appendix by the translator. Bombay, 1847. 50 leaves, numbered by Burton: [title-page], i–iv, 1/2, 3–46. Limp navy-blue wrappers.

104*b*
— typescript. [i], 78 leaves.

105
Translations from the poems of Hāfiz of Shīrāz. [*c.* 1875]. 9 leaves, and 1 leaf (inside upper wrapper). Holograph notebook; monogram IB, on upper wrapper.
Inserted in *Háfiz of Shīrāz: selections from his poems;* translated from the Persian by Herman Bicknell (*see* 880). *See also* 876–9.

106
To the Gold Coast for gold, or vingt ans après: a personal narrative. [1882]. [675] leaves. front. (holograph), [1] printed leaf. Leaves [454–61] manuscript by Edward L. McCarthy. *See also* 88.

107
The book of the sword (continuation) – 5 boxes:

Box 1: The secrets of the sword. 256 leaves headed by Burton: First evening – Ninth evening [rearranged by A. Forbes Sieveking] – 46 leaves headed by Burton: Bibliography of the sword; Senese pp. 67–70; Books (mostly Italian) on the duello; The fencing school (ancient and modern) – [Eight queries on Japanese swords], 2 leaves.

h j
Box 2: Miscellaneous, 1: 141 leaves of miscellaneous notes (inserted autograph letter signed: George Paget, Scutari, 19 April 1884) – autograph letters and postcards signed: A. Maclaren, Oxford, n.d. – Bernard Quaritch, London, 27 Jan. 1874 – Punch Office, [London], 5 Feb. 1875 [to IB] – D. E. Colnaghi, Florence, 29 March 1875 – William Tinsley, [London], 3 June 1875 – for Herbert & Co., London, 30 Aug. 1875 – E. W. Brock, Trieste, 16 Oct. 1875 – R. B. Joyner, [n.p.], 27 Feb. 1876 – H. Schültz Wilson, London, 27 May, 20 Sept., 3 and 16 Dec. 1876, 23 March 1877 – H. Burgess, Hydrabad Sind, 4 June 1876 – M. G. Mattilich, Trieste, 31 Oct. 1876 – for Chatto & Windus, and Andrew Chatto, London, 23 Nov. 1876, 5 April 1877 – [C.] Guillain, Cairo, 23 July 1877 – A. Lane Fox or A. Pitt Rivers, Guildford, 27 July 1877, [London], Salisbury, 21 March 1883, 4 March 1884 – Casafiel[?], Barcelona, 14 March, 27 April 1879 – Juan Eugenio Hartzenbusch, Madrid, 23 April 1879 – for *Iron,* London, 12 April, 14 May 1881 – H. Yule, [London], 11 Sept. 1882 – A. H.

Sayce, Bath, 6 Oct. 1882 – Joseph Grego, [London], 21 Sept. 1883 – Fred. W. Foster, London, 31 Jan., 9 May 1884 – Rich. S. Charnock, R. S. Charnock, or in Arabic, Athens, Nice, New Thornton Heath [etc.], 20 Feb., 14 March, 2, 8, 19 May, [23 June, 4, 9 July], 7 Aug., [25 Sept.] 1884, 6 March, [21 April] 1885 – W. Wareing Faulder, Manchester, 4 March 1884 – Alexder. Gutton – Walter Gregor, Fraserburgh, 8 Aug. 1884 – Edwin H. Baverstock, London, 3 Dec. 1884 – Foelin Tribolati, Pisa, 4 March 1885 – signed: Boy No. 2 [to Boy No. 1], London, [n.d.] – F. P. Verney, Claydon, Bucks, 28 Nov. [188-] – 3 autograph letters [not to Burton] – miscellaneous manuscript notes (32 by R. S. Charnock) – for pamphlet by Urbani de Gheltoff [*see* 1481·1] – printed matter.

h j

Box 3: Miscellaneous 2: 114 leaves, 12 sketches – autograph letters signed: J. Mayall, jr, [London], 30 Jan., 24 Feb. 1879; F.K.J. Shenton, Sydenham, Crystal Palace Company, 6 Jan. 1877 (to Georgiana Stisted); Percy Fielding, Auberge d'Aragon, Malta, 13 Oct. 1881; Yacoub Artin Bey, Cairo, 21 March 1881; from William Clowes, 4 Aug. 1876; C. Blom, [London], 5 July 1877; Abd-al-Shaikan, London, 9 May 1879; F. W. Simms, Cairo, 16 July 1876; M. G. Mattilich, Trieste, 19 Aug. 1876 – translation by M. G. Mattilich of *A new system of sword exercise for infantry,* 44 manuscript leaves, corrected by Burton, with Burton's letter to the translator on leaf 1 – certificate 'Brevet de pointe', Ecole d'Escrime, Paris (undated and unsigned) – pamphlet [*see* 1416] – printed matter.

h j

Box 4: Miscellaneous 3: autograph letters signed: C. Kraus, Mexico, 29 Feb. 1868; Chas. S. Dundas, Tenerife, 9 May 1882; in Arabic: Abd-al-Shaikan, London, 27 July 1873; Bernard Quaritch, London, 16 Feb. 1874; I. W. Grundy (?) London, 10 May – (to IB); Edmund Yates, London, 6 Feb. 1875 (to IB); Violet Greville, [London], 23 April [no year] (to IB); J. W. Russell (?) [Aberdeen], 31 Aug. 1875 (to IB); John Kirby, Zanzibar, 7 Aug. 1873; John Latham, [London], 25 May, 14 Sept. 1876 (to H. S. Wilson) – pamphlets [*see* 1411, 1453-4, 1485] – manuscript notes and sketches by Frank Hutton – printed matter.

j

Box 5: Miscellaneous 4: autograph letters signed: C.W.G., Edinburgh, 8 July, 23 Sept. [1875?]; in Arabic: Abd-al-Shaikan, London – 1 holograph leaf (on illus.) – manuscript note from Alex. Fergusson, Edinburgh, 26 April 1884 – notes, 19 leaves, signed: R. S. Charnock, New Thornton Heath, 12 July 1884.

108 f h j

Selected system of fencing, compiled by R. F. Burton. [126] leaves.
Insertions: autograph letter signed: P. P. Cautley, Trieste, 8 May 1885 – printed matter.

See also: holograph sketches in 10, 13, 18, 50, 54-5, 60, 108 *et passim*, and holograph leaves inserted in 3, 10-11, 13, 26, 33, 50-1, 54-5, 60, 79*b*, 88, 89*c-d*, 90, 91*a*, 94*a*, 112 *et passim.*

MISCELLANEA

109

Société de Geographie, Paris. Prix annuel pour l'année 1857, aux Capitaines Richard F. Burton et J. H. Speke pour leur exploration des Grands lacs de l'Afrique orientale. Certificate dated Paris, le 20 avril 1860. 1 sheet.

110–122 Works by or attributed to Lady Burton

110 l

Scenes in Tenerife. [after 1863]. 63-70, 237-45, 339-45, 554-60 pp. Reprinted from an unidentified journal.
Contents: 1, Santa Cruz to Oratava – 2, A climb up the peak – 3, On the peak, and down again – 4, Some account of the Guanches.
Probably by IB who was in Tenerife in March 1863 (*see* 120 for *The Life,* Vol. 1, pp. 380-1).

111 l

The inner life of Syria, Palestine, and the Holy land, from my private journal, by Isabel Burton. London, Henry S. King, 1875. 2 vols. fronts. (ports.), plate, map.

112 d h l

AEI: Arabia Egypt India, a narrative of travel, by Isabel Burton. London, Belfast, William Mullan, 1879. viii, 488 pp., plates, maps.
Burton's copy: pp. 448-53 heavily annotated by him.

g j l

Another copy, H. R. Arundell's, *i.e.* IB's father to whom the book is dedicated.
Insertions: autograph letters signed: George Holme, or G. Holme, Southport, 6 Dec. 1886 and 20 June 1887 – 2 pp. manuscript extracts from pp. 394-5.

113

Prevention of cruelty, and anti-vivisection . . . dedicated to my fellow-workers in the great cause of humanity. London, Belfast, William Mullan, 1879. 32 pp., illus.
Preface: 'I am enabled to abstract from my new book – "A.E.I.: Arabia, Egypt, and India" – all my remarks on Animals, Prevention of Cruelty, and Anti-Vivisection, and to form them into a pamphlet'. [*see* 219].

114

Alcune parole, sulla necessitá di fondare un seminario in Trieste; indirizzate da Isabel Burton, ai Triestini. [London, Unwin Brothers, printers, 188-]. 12 pp. [*see* 219].

115

Mrs Richard Burton's annual fête and distribution of prizes [in Trieste]. London, T. Vickers Wood, printer, [188-]. An offprint of 20 pp. Reprinted from an unidentified journal. [*see* 219].

116

La solenne distribuzione dei premi della Società Zoofila Triestina [address in Italian]. From *Bollettino della Società Zoofila Triestina,* 31 dicembre 1877, Anno 15, Punt. IV, pp. 53-5, and 15 dicembre 1879, Anno 17, Punt. III/IV, pp. 33-6. [*see* 219].

117 j l

Sketches of Indian life and character for boys by Aunt Puss (Isabel Burton). London, Belfast, William Mullan, 1880. xii pp.
Preface only; p. viii: 'a series of character-sketches and ceremonies culled from all Captain Burton's works on India, and my one book written whilst there.' The book does not appear to have been published.
Insertion: holograph note to Martha Brown, signed Isabel Burton, 26 Feb. 1879, with reply dated 21 March, attached. [*see* 219].

118

The reviewer reviewed, by Isabel Burton. [London?], privately printed, [1881]. An offprint of 20 pp. Reprinted from *Camoens: his life and his Lusiads,* Vol. 2 [*i.e. The Lusiads,* Vol. 4], pp. [709]-27.
Five other copies (4 inside lower cover; for the 5th *see* 219).

119

Mrs Richard Burton renders an account of her stewardship to the donors to her animal fund at Trieste, fête of 17th June, 1883. [London, Ballantyne Press, 1883]. 15 pp. [*see* 219].

120

The life of Captain Sir Richd. F. Burton, K.C.M.G., F.R.G.S., by his wife, Isabel Burton. London, Chapman & Hall, [1893]. 2 vols. fronts. (ports), illus., plates. At head of title: [*in red:*] Presented by desire of / the late Isabel Lady Burton. Lettered on both upper covers: [illustration of a tombstone with lettering: + Richard Burton / 20th October 1890 / R.I.P.; with illus. of a dog] / Presented by / His Wife, Isabel Burton, / in affectionate memory of / Sir Richard Burton.

121

Opinions of the Press and of scholars on the "Arabian nights." [London, 1893?]. 21 pp. Reprinted from her *The Life,* Vol. 2, Appendix I, pp. 617–28 with some re-arrangement of the order.

122

Proofs and cuttings from newspapers [*see* 219]:

a. Mrs Burton on cruelty to animals. Proof of letter to the *Morning Post,* dated Trieste, 6 Sept. [1881].

b. Birds killing their young, dated Trieste, 26 July.

c. Fidelity and instinct in dogs, dated Trieste, 20 July 1884/5? (*see* her *The Life,* Vol. 2, p. 278).

d. Instinct in birds, dated Cairo, May 1, 1878.

e. Contra coloro che maltrattano le bestie. [188–].

123-72 General

123 d
Anonymous. Du groupement des peuples et de l'hégémonie universelle. Paris, [*c.* 1875]. 3-14 pp. Reprinted from an unidentified journal. [*see* p.].

124
— The manual of heraldry. London, H. Salt Heraldic Office, [1861]. 132 pp., front., plate, illus.

125 a e h
Accademia di Commercio e Nautica in Trieste. Prospetto degli studj . . . per l'anno scolastico 1870-71, 1871-72. Trieste, Lodovico Herrmanstorfer, 1871-2. 2 parts. [*see* p.].

126 d
Ahmad ibn Muhammad, called Ibn Khallikān. Ibn Khallikan's Biographical dictionary translated from the Arabic by Mac Guckin de Slane. Paris, printed for the Oriental Translation Fund of Great Britain and Ireland, 1842-71. 4 vols.

127 a i
Allen, E. H. Hodges against Chanot: being the history of a celebrated case. London, Mitchell & Hughes, printer, for the author, 1883. 88 pp. No. 15 of an edition limited to 90. (26 March 1883). [*see* p.].

128
Beatson, W. F. Lord Stratford de Redcliffe, the War Department, and the Bashi Buzouks. London, W. Clowes, 1856. 111, 12 pp. [*see* 224].

129
— Remarks on the evidence given before the Court of Inquiry, on charges preferred against Major-General Shirley. London, W. Clowes, 1857. 12 pp. [*see* 224].

130 d
— The War Department and the Bashi Buzouks. London, W. Clowes, printer, 1856. 119 pp. [*see* 237].

131 a
[Bianconi, collectors]. Catalogo delle pitture e sculture possedute dalla famiglia Bianconi in Bologna. Bologna, S. Tommaso d'Aquino, printer, 1854. 18 pp. [*see* p.].

132
Brande, W. T. and G. W. Cox, eds. A dictionary of science, literature, & art . . . new edition, revised. London, Longmans, Green, 1875. 3 vols. illus.

133 a d
Brewer, E. C. Dictionary of phrase and fable, giving the derivation, source, or origin of common phrases, allusions, and words that have a tale to tell; third edition. London, [etc.], Cassell, Petter, & Galpin, [1872]. xii, 984 pp., illus.

134 k
[Bunce, O. B.]. Don't: a manual of mistakes and improprieties more or less prevalent in conduct and speech . . . by Censor. London, Griffith & Farran, 1884. 77 pp.

135 a d
Chambers, W. and R. Chambers, eds. Chambers's information for the people; new edition. London, Edinburgh, W. & R. Chambers, [1857]. 2 vols. (in 1) front., illus., maps.

136 a d
[Chandler, H. W.]. On lending Bodleian books and manuscripts. [Oxford, Baxter, printer, 1886]. 31 pp. [*see* p.].

137 a d
— [On book-lending as practised at the Bodleian Library. Oxford, Baxter, printer, 1886]. 30 pp.
138 l
Chompré, P. Dictionnaire abrégé de la fable, pour l'intelligence des poëtes, des tableaux et des statues dont les sujets sont tirés de l'histoire poëtique; quinzième édition. Riom, Imprimerie de Landriot, [etc.], [*c.* 1800]. 456 pp.
139
Crystal Palace. Grand fête in honour of His Highness the Sultan of Zanzibar, Saturday, June 19th, 1875: programme & book of words. [London], 1875. [16]pp. [*see* p.].
140 a d
Encyclopedia historica, politica, geographica e commercial . . . Angra do Heroismo, Imprensa de J. J. Soares, 1840. 278, 94 pp.
141 l
Gobat, S. and others. Brief des Bischofs Samual Gobat . . . Berlin, 1861. 33–64 pp. From *Neueeste Nachrichten aus dem Morgenlande,* Jahrgang 5, No. 18. [*see* 225].
142
Grant, J. The newspaper press: its origin-progress-and present position. . . London, Tinsley Brothers, 1871. 2 vols.
143
[Great Britain. Parliament]. Humble adresse de la Chambre des Communes a Son Altesse Royale Le Prince Regent, Mardi, le 3 de mai, sur le commerce des esclaves en Afrique et la response de Son Altesse Royale, vendredi, le 3 de juin, 1814. [Londres, H. Hay, printer], 1814. 7 pp. [*see* p.].
144 a e i
Hortis, A. Giovanni Boccacci . . . Ambasciatore in Avignone e pileo da prata . . . proposto da' Fiorentini a patriarca di Aquileia. Trieste, L. Herrmanstorfer, 1875, 83 pp.
145
Howard, J. J. ed. Miscellanea genealogica et heraldica [advertisement only]. London, Hamilton, Adams, [1866]. [4] pp. [*see* 257].
146 a e
Imprimerie arménienne de Saint-Lazare. Catalogue des livres. Venise, Etablissement des Mékhitharistes, 1869. 34 pp. [*see* 259].
147
Janus, pseud. Why women cannot be turned into men. Edinburgh, London, William Blackwood, 1872. 24 pp. [*see* 223].
148
Lacordaire, H. D. Discours de réception à l'Académie française. Paris, Librairie de Mme Ve Poussielgue-Rusand, 1861. 48 pp. [*see* 224].
149 e
Laemmert, E. and H. Laemmert. Catalogo dos livros em Portuguez publicados e a venda no Livraria Universial dos editores-proprietarios. . . Rio de Janeiro, Typographia Universal de Laemmert, [1848]. 64 pp. [*see* 242].

150
Mahmūd, Bey. Mémoire sur le calendrier arabe avant l'Islamisme, et sur la naissance et l'âge du Prophète Mohammad, par Mahmoud Effendi. Paris, Imprimerie Impériale, 1858. [iv], 84 pp. Reprinted from *Journal asiatique,* Series 5, Vol. 11. [*see* p.].
151
Meriton, H. and J. Rogers. A circumstantial narrative of the loss of the Halsewell East-Indiaman Capt. Richard Pierce which was unfortunately wrecked at Seacombe in the Isle of Purbeck . . . Hamburg, J. G. Herold, 1794. [xiv], 218 pp. In English and German.
152 d l
Mill, J. S. Considerations on representative government. London, Parker, Son and Bourn, 1861. viii, 340 pp.
153 d
— Dissertations and discussions political, philosophical, and historical reprinted chiefly from the Edinburgh and Westminster reviews. London, John W. Parker, 1859. 2 vols.
154
— Essays on some unsettled questions of political economy. London, John W. Parker, 1844. vii, 164 pp.

155
Oriental Translation Committee. Report, prospectus, and publications . . . 1861. London, Harrison, printer, 1861. 12 pp. [*see* 228].
156 i
Oxford. University. Indian Institute. An account of the circumstances which have led to its establishment, a description of its aims and objects. Oxford, Horace Hart, printer to the University, 1884. 20, 8 pp., front. (Richard Burton Esq. as a subscriber). [*see* p.].
157
Penley, A. The elements of perspective: illustrated by numerous examples and diagrams; third edition. London, Winsor & Newton, 1851. iv, 60, 24 pp., illus., plate. [*see* 223].
158 a d
Phillips, Sir R. A million of facts, of correct data, and elementary constants, in the entire circle of the sciences . . . stereotyped edition. London, Darton, 1859. xlix pp., columns 1128, [34] pp., front., illus.
159 a e
Philo-Israel, pseud. The British constitution and Israel's policy, foreign and domestic; followed by "A necessary consequence" by L. P. London, W. H. Guest, 1880. 20 pp. Reprinted from the *Banner of Israel.* [*see* p.].
160
Polybius. The general history of Polybius in five books; translated from the Greek by Mr Hampton. London, printed by H. S. Woodfall for J. Dodsley; and printed for T. Davies, 1772–3. 4 vols. fronts (maps). Vols. 1–2 are of the 3rd edition; wanting Vol. 3.
161 a d
Riley, H. T. ed. A dictionary of Latin and Greek quotations, proverbs, maxims and mottos, classical and mediaeval. London, George Bell, 1878. vi, 622 pp.
162 l
Robinson, J. An easy grammar of history, ancient and modern . . . with questions and exercises . . . fourth edition. London, Richard Phillips, 1808. 156 pp.
163 i
Royal Asiatic Society. Bombay Branch. General catalogue of the library of the Bombay Branch of the Royal Asiatic Society corrected up to 31st December 1873. Bombay, [Byculla, Education Society's Press, printer], 1875. 854 pp. (J. A. Shepherd, 1874).
164
Rumsey, A. Will-making made safe and easy: an aid to testators. London, John Hogg, 1888. 140 pp.
165
Sette of odd volumes, The. London, imprynted by Bro^{r} C.W.H. Wyman, [etc.], 1880–7. 10 vols. fronts, plates.
166 a
Smith, W. ed. Dictionary of Greek and Roman biography and mythology. London, Taylor and Walton, 1844. 3 vols.
167 d k
Smythe, P. E. F. W., 8th Viscount Strangford. A selection from the writings of Viscount Strangford on political, geographical, and social subjects; edited by Viscountess Strangford. London, Richard Bentley, 1869. 2 vols. fronts.
168 h k
Sterni, A. Società, domestica, civile e religiosa al secolo decimonono. Piacenza, Tipografia di A. del Maino, 1878. xxi, 592 pp.
169 i
Walford, C. The destruction of Libraries by fire considered practically and historically. London, printed at the Chiswick Press, 1880. 36 pp. Reprinted from *Transactions and Proceedings of the Conference of Librarians,* Manchester, 1879.
170
Wharton, J. J. S. The law-lexicon, or dictionary of jurisprudence; second edition. London, V. & R. Stevens and G. S. Norton, 1860. xii, 776 pp., illus.
171 i
York Gate Library. Catalogue of the York Gate geographical and colonial library. London, John Murray, 1882. lxi, 134 pp.
172 e
— Catalogue of the York Gate library formed by Mr S. William Silver: an index to the

literature of geography, maritime and inland discovery, commerce and colonisation, by Edward Augustus Petherick; second edition. London, John Murray, 1886. cxxxii, 336 pp., plates.

173-216 Periodicals, Congress Reports

173
African-Aid Society, London, 1862. First report from July, 1860 to the 31st March, 1862. [*see* 238].

174 e
American Anti-Slavery Society, New York, 1860. Annual report . . . for the year ending May 1, 1859. [*see* 248].

175 i
An-nahlah, the bee: an illustrated and western periodical, London, 1877. Vol. 1, Nos. 2–4; and prospectus. [*see* p.].

176 e
Anthropological Institute of Great Britain and Ireland, London, 1879–83. Journal. Vol. 9, no. 2; vol. 12, no. 4; vol. 13, no. 1.

177 d
Anthropological Society of London, London, 1863/4-1867/9. Memoirs. Vols. 1-3.

178
Archivo litterario, Sao Paulo, julho, setembro 1865. Nos. 1, 3. [*see* 1801].

179
Brazil. Instituto historico, geographico e ethnographico do Brasil. Rio de Janeiro, 1864. Revista trimensal. Vol. 27, Part 1. [*see* 253].

180
British & Foreign Bible Society, London, 1812–27. Reports, 8th (5 copies); 21st (2 copies) and Report with appendix; 23rd. [*see* p.].

181 a
Calcutta Society for the Prevention of Cruelty to Animals, Calcutta, 1876. Report for the years 1874–6. [*see* p.].

182
Chile. Anales de la Universidad de Chile, Santiago, marzo 1862. Vol. 20, no. 3. [*see* 1801].

183 a d h i
Congrès international d'Anthropologie et d'Archéologie préhistoriques, Bologne, 1873. Compte rendu de la cinquième session à Boulogne 1871. xxxi, 543 pp., illus., plates (some col.) (Prof. [G. G.] Bianconi).

184
De Bow's review, [New Orleans], May 1859. N.S. Vol. 1, No. 5, pp. 487–610. [*see* 245].

185 a
Edinburgh review: or critical journal, The, London, Edinburgh, 1888. No. 343. Pp. 92–119 'The Heptameron of Marquerite of Navarre'.

186
Egypt. Arkān Ḥarb al-Jaish al-Misri. Jarīdat arkān ḥarb al-jaish al-Miṣrī [Bulletin of the Staff of the Egyptian Army; Arabic text]. Cairo, Maṭba't 'Umūm Arkān Ḥarb, 1294 A.H. [1876–7 A.D.]. Year 3, Nos. 1, 4, 6. [*see* p.].

187
[Egyptian Geographical Society]. Nashrat al-Jam'iyat al-Jughrāfiyat al-Khadiwiyah [Journal of the Egyptian Geographical Society. No. 8; Arabic text]. Būlāq, 1886 A.D. [*see* p.].

188
Englishwoman's review of social and industrial questions, London, 1875. N.S. Vol. 6, no. 2. [*see* p.].

189 a b
Ethnological Society of London, London, 1850–6. Journal. Vols. 2–4.

190 a d
— 1863, 1866–7. Transactions. N.S. Vols. 2, 4–5.

191
Ethnologische Mitteilungen aus Ungarn, Budapest, 1888. Vol. 1, no. 2 (1887–8). [*see* p.].

192 a d i l
Folk-lore journal, The, London, 1884–7. Vol. 2, nos. 3–4; Vol. 5, nos. 2–3 (in 1). (W.A.C. *i.e.* W. A. Clouston).

193 a d h
Folk-lore record, The, London, 1880–1. Vol. 3, nos. 1–2.

194
Future: a monthly journal of predictive science, The, London, October 1893. Vol. 2, no. 221. [*see* p.].

195
Harper's new monthly magazine, New York, 1859–60. Vol. 18, no. 104 (Jan. 1859), pp. 145–288; Vol. 21, nos. 125–6 (Oct.–Nov. 1860), pp. 577–864. [*see* 243].

196 d
Hutchings' California magazine, [San Francisco], April 1860. Vol. 4, no. 10. Pp. [433]–[40] 'Notes and sketches of the Washoe country'. [*see* 246].

197 a d
Institut égyptien, Alexandrie, Caire, 1874/5–80. Bulletin, No. 13; Serie 2, No. 1.

198 a d h i l
Istria, L', Trieste, 1846–52. Vols. 1–7 (in 4). Wanting Vol. 4, nos. 6, 26. (Gaetano J. Merlatoff).

199
Missionary herald, The, Boston, Massachusetts, March 1826–Oct. 1827. Vol. 22, nos. 3–8, 10–12; Vol. 23, nos. 1–5, 7–10. [*see* p.].

200
Missionary register, containing the principal transactions of the various institutions for propagating the gospel. London, Jan.–Feb. 1828. [*see* p.].

201 d
Museo Canario, El, Las Palmas. Revista quincenal órgano de la Sociedad del Mismo Nombre. 22 Jan. 1882. Vol. 4, no. 46. [*see* p.].

202 d
North American review, Boston, January 1867. Vol. 104, no. 214, pp. 1–30, 122–42. Contents: [Reviews of] *A discourse of Virginia,* by E. M. Wingfield; and *A true relation of Virginia,* by J. Smith – *The Albert Nyanza,* by S. W. Baker; *Journal of the discovery of the sources of the Nile,* by J. H. Speke; and *The sources of the Nile,* by C. T. Beke. [*see* 218].

203
Pennsylvania Hospital. Report of the Board of Managers . . . annual meeting, held fifth month 7th, 1860. Philadelphia, Collins, printer, 1860. 32 pp., front. [*see* 245].

204 a e
Royal Asiatic Society of Great Britain and Ireland, London, 1881. Journal. N.S. Vol. 13, no. 1.

205 a d h
— North China Branch, Shanghai, 1877. Journal. N.S. No. 11. Pp. 143–84 'Chinese eunuchs', by G. Carter Stent.

206 a d
Royal Geographical Society, London, 1831–60. Journal. Vols. 1–30; and Index to Vols. 1–20.

207
— London, 1867. Proceedings, Session 1866–67. Pp. 231–46 'On Dr. Livingstone's last journey and the probable ultimate sources of the Nile' by Alex. Geo. Findlay. [*see* 228]. *See also* 1571.

208
— London, 1886. Supplementary papers. Vol. 1.

209
Royal Society for the Protection of Animals in Florence, Florence, 1889. [Annual report], 1888/9. [*see* p.].

210 a d
Society of Biblical Archaeology, London, 1873–80. Transactions. Vol. 2, no. 1; Vol. 4,

no. 2; Vol. 5, nos. 1–2; Vol. 6, no. 2; Vol. 7, no. 1.

211 d k

Spectator, The, London, J. and R. Tonson and S. Draper, [1711–14]. Vols. 1–8, nos. 1–635.

212

Symons's monthly meteorological magazine, London, Sept. 1871. No. 68. [*see* 257].

213

Tweddell's advertising sheet, Stokesley, [1869]. No. 12. [*see* 257].

214

United Presbyterian Church. Missionary record. Edinburgh, 1 Nov. 1861. Vol. 16. Pp. 196–9 'Old Calabar, Duke Town: cruel and sanguinary deeds', anonymous. [*see* 238].

215

Vanity fair. London, 1869–70. 3 vols of cuttings. [*see* q.].

216

Zagreb. Popis Arkeologičkoga Odjela Nar. Zem. Muzeja u Zagrebu. Vols. 1–2 (1899–90).

217-61 Pamphlets Bound by Burton

The following:

(a) carry his visiting-card or book-plate:
  (i) *book-plate:* 224, 243
  (ii) *visiting-card:* 218, 220-1, 238, 243, 246, 257
(b) are lettered in gold on upper cover:
  (i) PAMPHLETS / [*in black letter:*] Richard F. Burton [*or*] Isabel Burton: 219-21
  (ii) [*in black letter:*] Ober Ammergau / Isabel Burton: 235

217 RFB 1
218 RFB 2
219 RFB 3
220 RFB 4
221 RFB 5
222 General 1838
223 General 1857
224 General 1860
225 General 1861
226 General 1863
227 General 1864
228 General 1869
229 General 1871
230 General 1873
231 Linguistics – Africa 1844
232 Linguistics – Africa 1848
233 Linguistics – Europe 1858
234 Religion 1857
235 Religion 1870
236 Sword 1854
237 Africa 1851
238 Africa 1858
239 Africa 1859
240 Africa 1859
241 Africa 1864
242 America 1842
243 America ?1856
244 America 1857
245 America 1857
246 America 1858
247 America 1859
248 America 1863
249 America 1864
250 America 1866
251 America 1866
252 America 1867
253 America 1868
254 America 1868
255 America 1872

256 Asia 1857
257 Asia 1863
258 Asia 1871
259 Asia 1872
260 Asia 1872
261 Europe 1822

262–303 Anthropology, Archaeology, Folklore

262 a d
Anonymous. The slavery quarrel: with plans and prospects of reconciliation, by a Poor Peacemaker. London, Robert Hardwicke, 1863. [ii], 51 pp. [*see* 226].

263 i
Beke, C. T. Views in ethnography, the classification of languages, the progress of civilization, and the natural history of man. London, Taylor & Francis, 1863. 16 pp. [*see* 241].

264 d
Blumenbach, J. F. The anthropological treatises of Johann Friedrich Blumenbach . . . with memoirs of him by Marx and Flourens, and an account of his anthropological museum by Professor R. Wagner, and the inaugural dissertation of John Hunter, M.D. On the varieties of man; translated and edited from the Latin, German and French originals, by Thomas Bendyshe. London, Longman, Green for the Anthropological Society, 1865. xvi, 406 pp., plates.
Another copy.

265 d h
Bolton, H. C. The counting-out rhymes of children: their antiquity, origin, and wide distribution: a study in folk-lore. London, Elliot Stock, 1888. xi, 123 pp.

266 a d
Brantôme, P. de B. Vies des dames galantes; nouvelle édition. Paris, Garnier Frères, 1872. [iii], 390 pp.

267
British Museum. Guide to the Christy collection of prehistoric antiquities and ethnography. London, Trustees of the British Museum, 1868. 24 pp. [*see* 223].

268 a d
Broca, P. On the phenomena of hybridity in the genus homo; edited . . . by C. Carter Blake. London, Longman, Green, [etc.] for the Anthropological Society, 1864. xiv, 120 pp.

269 a
Cairnes, J. E. and G. McHenry. The Southern Confederacy and the African slave trade: the correspondence between Professor Cairnes, A. M., and George McHenry; with an introduction and notes by the Rev. George B. Wheeler. Dublin, McGlashan & Gill, 1883. xxviii, 61 pp. [*see* 226].

270 a d j
Clarke, H. The Khita and Khitha-Peruvian epoch: Khita, Hamath. London, N. Trübner, 1877. vii, 88 pp., tables. (Hyde Clarke; 27 Dec. 1880). [*see* p.].

271
— On the propagation of mining and metallurgy. [London, 1868]. 4 pp. Reprinted from the *Transactions of the Ethnological Society,* N.S., Vol. 6. [*see* 256].

272 a d
Clermont-Ganneau, C. Horus et Saint Georges d'après un bas-relief inédit du Louvre: notes d'archéologie orientale et de mythologie sémitique. Paris, 1877. 51 pp. illus., plate. Reprinted from *Revue archéologique,* 1876. [*see* p.].

273 a d
Cuvier, G. L. C. F. D. de, Baron. The animal kingdom, arranged after its organization, forming a natural history of animals . . . translated and adapted to the present state of science; a new edition, with additions by W. B. Carpenter . . . and J. O. Westwood . . . London, Wm. S. Orr, 1849. x, 718 pp., front. (col.), illus., plates.

274 i
Davis, J. B. The neanderthal skull: its peculiar conformation explained anatomically. London, Taylor & Francis, printer, 1864. 16 pp., illus. [*see* 228].
275 i
— Note on the distortions which present themselves in the crania of the ancient Britons. [Dublin, London], 1862. 7 pp., illus. Reprinted from the *Natural history review*, July 1862. [*see* 228].
276 i l
— Sur les déformations plastiques du crâne. Paris, 1862. 379–90 pp., plates. From *Mémoires de la Société d'Anthropologie de Paris*, Series 1, Vol. 1. [*see* 222].
277 i l
— Thesaurus craniorum: catalogue of the skulls of the various races of man . . . London, printed for the subscribers, 1867. xvii, 374 pp., illus., plates, tables.
278 d h i j
Day, St J. V. The prehistoric use of iron and steel, with observations in certain matters ancillary thereto. London, Trübner, 1877. xxiv, 278 pp., front., illus., plates, tables. (St John V. Day; 7 Oct. 1881; Glasgow, 7 Oct. 1881).
279 i l
Donovan, C. The Ethnological Society and phrenology: a paper, entitled 'Physiognomy popular and scientific', read . . . 24th May 1864 before the Ethnological Society. London, William Tweedie, 1864. 24 pp. [*see* 228].
280
Great Britain. Parliament. Correspondence with the British Commissioners, at Sierra Leone, Havana, the Cape of Good Hope, Loanda, and New York; and reports from British Vice-Admiralty courts, and from British naval officers, relating to the slave trade, from January 1 to December 31, 1866; presented to both Houses of Parliament. London, Harrison, printer, [1867]. vi, 113 pp., tables. At head of title: Class A. [*see* q.].
281 d
— — Correspondence with British Ministers and agents in foreign countries, and with foreign ministers in England, relating to the slave trade from January 1 to December 31, 1863; presented to both Houses of Parliament. London, Harrison, printer, [1864]. xi, 199 pp., tables. At head of title: Class B. [*see* q.].
282 a d
Hopkins, J. H. The Bible view of American slavery: a letter from the Bishop of Vermont. London, Saunder, Otley, 1863. 15 pp. Reprinted from the *Philadelphia Mercury*, 11 Oct. 1863. [*see* 226].
283
Hunt, J. Anniversary address, delivered before the Anthropological Society of London, January 5th, 1864 [and First annual report 1863]. London, Trübner, 1864. 32 pp. Preceded by an outline of the Society's objectives, 4 pp. [*see* 226].
284 a d
— Introductory address on the study of anthropology, delivered before the Anthropological Society of London, February 24th, 1863. London, 1863. 20 pp. [*see* 226].
285
— [Letter to Dr John Beddoe, President of the Anthropological Society of London. London, printed for private circulation, 1869]. 8 pp. [*see* 223].
286 d i
— On ethno-climatology: or the acclimatization of man. [London], 1862. 31 pp. Reprinted from the *Transactions of the Ethnological Society*, N.S., Vol. 2. [*see* 228].
287 i
— On the Negro's place in nature . . . read before the Anthropological Society of London. London, Trübner for the Anthropological Society, 1863. viii, 60 pp. Pp. [v]-viii dedicatory letter to Burton. [*see* 241].
h
Another copy; *Circular*, No. 4 of the Anthropological Society inserted. [*see* p.].
Third copy [*see* 226].
288
Lane Fox, A. H. afterwards Pitt Rivers. Catalogue of the anthropological collection lent . . . for exhibition in the Bethnal Green Branch of the South Kensington Museum, June 1874. London, George E. Eyre and William Spottiswoode for H.M.S.O., 1877. xvi, 184 pp., illus., plates. imperfect. [*see* p.].
289
Mantica, N. Raccolta di proverbi e dittati ippici. Udine, Tipografia del Patronato, 1883. 110 pp.

290 a d h
Mitchell, A. The past in the present: what is civilisation? Edinburgh, David Douglas, 1880. xvi, 354 pp., illus.

291
Much, M. Bericht über die Versammlung österreichischer Anthropologen und Urgeschichtsforscher am 28. und 29. Juli 1879 zu Laibach. Wien, Anthropologische Gesellschaft, 1880. 124 pp., illus. Reprinted from *Mittheilungen der anthropologischen Gesellschaft in Wien,* Vol. 10. [*see* p.].

292 d
Pouchet, G. The plurality of the human race; translated and edited (from the second edition), by Hugh J. C. Beavan. London, Longman, Green, [etc.] for the Anthropological Society, 1884. xiv, 158 pp.
Another copy.

293
Praeuscher, E. Nuova guida per il Museo anatomico, patologico ed etnologico. Trieste, G. Caprin, 1877. 50 pp. [*see* p.].

294 a d
Prichard, J. C. The natural history of man; comprising inquiries into the modifying influence of physical and moral agencies on the different tribes of the human family; third edition, enlarged. London, Hippolyte Bailliere, [etc.], 1848. xvii, 677 pp., front. (col.), illus., plates, tables.

295 a d g h j
Reade, W. The martyrdom of man; twelfth edition. London, Trübner, 1887. viii, 544 pp. (L. S. Metcalf, New York, 20 July [188-]).

296 d i
Rogers, E. T. Unpublished glass weights and measures. [London, 1878]. 15 pp., front. Wrapper dated 1877. Reprinted from the *Journal of the Royal Asiatic Society,* N.S. Vol. 10. [*see* p.].

297 f g k
Rollin, M. The ancient history of the Egyptians, Carthaginians, Assyrians, Babylonians . . . translated from the French. London, Henry G. Bohn, 1958. 2 vols. (in 1) front., maps.

298 i
Tylor, Sir E. B. On a method of investigating the development of institutions, applied to laws of marriage and descent. London, Harrison, printer, 1889. 245–69 pp. Reprinted from the *Journal of the Anthropological Institute,* Vol. 18. [*see* p.].

299 e
Vaux, W. S. W. On recent additions to the sculptures and antiquities of the British Museum. London, [1866]. 38 pp., plate. From the *Transactions of the Royal Society of Literature,* Series 2, Vol. 8, pp. 559–96. [*see* 241].

300 d
Vogt, C. Lectures on man: his place in creation, and in the history of the earth; edited by James Hunt. London, Longman, Green, [etc.] for the Anthropological Society, 1864. xxii, 476 pp., illus.
Another copy.

301 a d
Waitz, T. Introduction to anthropology; edited, with numerous additions by the author, from the first volume of "Anthropologie der Naturvölker" by J. Frederick Collingwood. London, Longman, Green [etc.] for the Anthropological Society, 1863. xvi, 404 pp.
Another copy.

302 l
West India Committee. Free labour and the slave trade. [London, Robert Barclay, printer, 1861]. 65 pp. [*see* 227].

303 e
— Papers relating to free labour and the slave trade: a corrected report of the debate in the House of Commons, on the 26th of February 1861, upon resolutions proposed by Mr Cave, the chairman of the West India Committee . . . London, Robert Barclay, 1861. 65 pp. [*see* 238].

304-71 Biography

304
Anonymous. Captain Richard Burton. [Rio de Janeiro, 1865]. 73-8 pp. Reprinted from the *Anglo-Brazilian Times,* Oct. 1865. [*see* p.].
305
— Captain Richard Burton. [London, *c.* 1872]. [1] p., port. Reprinted from an unidentified journal. [*see* p.].
306
— Le Général Comte Luigi Palma di Cesnola (Italie). [Paris, 1869]. 12 pp., front. (port.). Reprinted from *L'histoire générale des hommes XIXe siècle vivants ou morts de toutes les nations.* [*see* pq.].
307
— Memoir of Capt. Richard Burton (extracts from the press). [London, *c.* 1864]. 3 pp. 10 copies. [*see* p.].
308
— Sir Richard Francis Burton, K.C.M.G. [London], 1890. 8 pp. Reprinted from the *Proceedings of the Royal Geographical Society,* Vol. 12. [*see* p.].
309
— Sir Roderick-Impey Murchison, Baronet (de la Grande-Bretagne). Genève, [1866]. 3 pp., front. (port.). Reprinted from *L'histoire générale des hommes du XIXe siècle vivants ou morts de toutes les nations.* [*see* pq.].
310
[Anthropological Society of London]. Anthropological news. [London], 1868. 461-4 pp. From the *Anthropological review,* Vol. 6. Pp. 462-3 reference to Burton's annotation of Waitz's *Anthropology of primitive peoples,* Vol. 2 (*see* 22). [*see* p.].
311
— Farewell dinner to Captain Burton [on his departure for Santos, Brazil]. [London], 1865. 167-82 pp. From the *Anthropological review,* Vol. 3. [*see* p.].
312
— [Report of the meeting held on 20 December 1864, Captain R. F. Burton, Vice-President, in the chair. London, 1865]. xlvi-lxxiv pp., plate. From the *Journal of the Anthropological Society,* Vol. 3. [*see* p.].
313 k
Austin, L. F., pseud. Frederic Daly. Henry Irving in England and America, 1838-84, by Frederic Daly. London, T. Fisher Unwin, 1884. viii, 300 pp., front. (port.).
314 a
Barrett, J. O. The spiritual pilgrim: a biography of James M. Peebles; second edition. Boston, William White, [etc.], 1872. 303 pp., front. (port.).
315
[Bensilum, A. and G. Weiss]. Schiarimenti, Bensilium di Cairo e Giacomo Weiss, ottico di Baviera a Trieste di fronte all'opinione pubblica. Cairo, 1884. 36 pp. [*see* p.].
316 d i
Besant, Sir W. The life and achievements of Edward Henry Palmer; second edition. London, John Murray, 1883. xi, 430 pp., front. (port.).
317
Brabrook, Sir E. W. Sir R. F. Burton, K.C.M.G. [London], 1891. 295-8 pp. Reprinted from the *Journal of the Anthropological Institute;* Vol. 20. [*see* p.].
318 a d
Bruce, W. N. Life of General Sir Charles Napier, G.C.B. London, John Murray, 1885.

416 pp., front. (port.), illus., plates, maps.

319 k

Burke, E. A speech of Edmund Burke, Esq. at the Guildhall in Bristol, previous to the late election in that city, upon certain points relative to his parliamentary conduct; third edition. London, J. Dodsley, 1780. 68 pp.

320 a d j

Cantù, C. Lord Byron and his works: a biography and essay; edited, with notes and appendix, by A. Kinloch. London, George Redway, printed for the Editor by the Shropshire Guardian Newspaper Company, 1883. [viii], 81, xiii pp. (A. Kinloch, [London], 15 June 1883).

321 d

Churchill, C. H. The life of Abdel Kader, ex-Sultan of the Arabs of Algeria, written from his own dictation, and compiled from other authentic sources. London, Chapman & Hall, 1867. xvi, 331 pp.

322

Collette, C. H. "Luther vindicated". London, Bernard Quaritch, 1884. vi, 226 pp.

323

Custis, G. W. P. Recollections and private memoirs of Washington. Washington, D.C., William H. Moore, printer, 1859. 105 pp.

324

Dallas, R. C. Recollections of the life of Lord Byron, from the year 1808 to the end of 1814 . . . taken from authentic documents, in the possession of the author. London, printed for Charles Knight, 1824. xcvii, 344 pp., front.

325

Dezos de la Roquette, J. B. M. A. Notice sur la vie et les travaux de M. Pierre Daussy. Paris, Imprimerie de L. Martinet, 1861. 27 pp., front. (port.). [*see* 224].

326

— Notice sur la vie et les travaux de M. le Baron A. de Humboldt. Paris, Imprimerie de L. Martinet, 1860. [iv], 88 pp., fronts. (ports.), plate. [*see* 224].

327

Duval, J. Un ouvrier voyageur René Caillié. Paris, L. Hachette, 1867. 52 pp. [*see* p.].

328 h

Edwards, E. Biography of Captain Burton: taken from 'photographic portraits of men of eminence'; from 'Extracts from the Press', and 'The Anglo-Brazilian Times,' of October 1865. [London, 1865]. 8 pp. At head of title: Printed for private circulation. [*see* p.].

329

Ford, R. Memoir and military biography of His Grace the late Duke of Wellington. London, the proprietors, [Gallery of Illustration], 1852. [ii], 68 pp., illus.

330 l

Froulay, R. C. V. de, afterwards Marquise de Créquy. Souvenirs de la Marquise de Créquy de 1710 à 1803; nouvelle édition . . . revue, corrigée et augmentée. Paris, Garnier Frères, [1840]. 10 vols. (in 5), fronts.

331

Gill, A. Gambetta avec un portrait, et un autographe; troisième édition. Paris, Sandox et Thuillier, 1882. [iii], 542 pp., fronts.

332 d

Gordon, C. G. The journals of Major-Gen. C. G. Gordon, C.B., at Kartoum, printed from the original MSS.; introduction and notes, by A. Egmont Hake. London, Kegan Paul, Trench, 1885. lxvi, 587 pp., front. (port.), illus., maps.

333 l

Greville, C. C. F. The Greville memoirs: a journal of the reigns of King George IV. and King William IV.; edited by Henry Reeve; second edition. London, Longmans, Green, 1874. 3 vols.

334 i

Guillemine, C. Notice nécrologique sur M. le Mis de Compiègne. Le Caire, Typographie française Delbos-Demouret, 1877. 20 pp. At head of title: Société khédiviale de Géographie. [*see* p.].

335 d g h j

Hitchman, F. Richard F. Burton, K.C.M.G.: his early, private and public life, with an account of his travels and explorations. London, Sampson Low, Marston, [etc.], 1887. 2 vols. fronts., illus., plates. (Ja. Crowdy, [London], 7 May –).

336

Kandler, P. Una corona al defuntó Bar. [one] [Giovanni Guglielmo] Sartoria. Trieste,

[1871]. 3 pp. Reprinted from *Gazzetta di Trieste,* No. 249, October 1871. [*see* 2264].

337 k l

Laferté, V. Alexandre II., détails inédits sur sa vie intime et sa mort. Bâle, Genève, Lyon, G. Georg, 1882. 219 pp.

Another copy.

338 a d h

Lane-Poole, S. The life of the Right Honourable Stratford Canning, Viscount Stratford de Redcliffe . . . from his memoirs and private and official papers. London, New York, Longmans, Green, 1888. 2 vols. fronts. (ports.), plates.

339 k

[Lockhart, J. G.]. The history of Napoleon Buonaparte . . . Vol. 1. London, John Murray, 1829. viii, 303 pp., front., plates.

340 l

[Luciani, T.]. Petro Kandler. [Venezia], 1872. 23 pp. Reprinted from *Archivio veneto,* Vol. 3, No. 1. [*see* p.].

341

Lundie, R. H. Alexander Balfour: a memoir: fourth edition. London, James Nisbet; Liverpool, Philip, 1889, vii, 348 pp., front. (port.).

342 a d i

Mackay, G. E. Lord Byron at the Armenian convent. Venice, Office of the "Poliglotta". 1876. [ix], 102 pp. (Venice, 27 Dec. 1875). [*see* p.].

k

Another copy.

343 d

[Marchesetti, C.]. Discorso tenuto in occasione dello scoprimento del busto di Bartolommeo Biasoletto il XVIII maggio 1878. Trieste, Tipografia di L. Herrmanstorfer, 1878. 7 pp. [*see* p.].

344 k

Medwin, T. Journal of the conversations of Lord Byron, noted during a residence with His Lordship at Pisa, in the years 1821 and 1822 . . . Vol. II. Paris, L. Baudry, 1824. [ii], 264 pp.

345 a

Melli, S. R. Parole lette in onore dell'illustre Baronetto Sir Moses Montefiore. [Trieste, Morterra, printer], 1883. 8 pp. [*see* p.].

346

Millar, R. A statement or various acts of gross cruelty, – oppression – and injustice – done to me – Surgeon-Major Robert Millar, M.D., (late) of H.M.'s Bombay Army. [London, 1869]. 67 pp. [*see* p.].

347 i

Milnes, R. M., Baron Houghton. Some writings and speeches of Richard Monckton Milnes, Lord Houghton in the last year of his life, with a notice in memoriam by George Stovin Venables. London, privately printed at the Chiswick Press, 1888. [vii], 138 pp. (H. G. Galway, August 1888).

348

Moore, T. Memoirs of the life of the Right Honourable Richard Brinsley Sheridan. Vol. II. Paris, A. & W. Galignani, 1825. vi, 500 pp., plate.

349 d

Mury, P. Histoire de Gabriel Malagrida de la Compagnie de Jésus, l'Apôtre du Brésil au XVIIIe siècle . . . Paris, Charles Douniol, 1865. iv, 272 pp.

350

Oxenham, H. N. Memoir of Lieutenant Rudolph de Lisle, R.N. of the Royal Naval Brigade on the upper Nile; second edition. London, Chapman & Hall, 1886. xxiv, 296 pp., front. (port.), plates.

351

[Pagani, P. and others]. Onoranze a Niccolò Tommaseo per cura della colonia Dalmata dimorante in Trieste. Trieste, Tipografia del Lloyd Austro-Ungarico, 1874. 96 pp. [*see* 230].

352 a d h

Pereira da Silva, J. M. Manuel de moraes: chronica do seculo XVII. Rio de Janeiro, B. L. Garnier; Pariz, Augusto Durand, 1866. iv, 287 pp.

353 k

[Proyart, L. B.]. The life of Princess Louisa (Madame Louise), of France: daughter of Louis XV. King of France, a Carmelite nun; a new edition. Salisbury, J. Easton, printer for Mrs Elizabeth Macdonald, 1808. 2 vols., front. (port.).

354 k

[Ramsden, Lady G. ed.]. Letters of Lord St. Maur and Lord Edward St. Maur, 1846 to 1869. London, printed for private circulation, 1888. [vii], 444 pp.

355 k l

[Richards, A. B. and A. Wilson]. A short sketch of the career of Captain Richard F. Burton, collected from "Men of eminence;" from Captain and Mrs. Burtons own works; from the Press, from personal knowledge . . . by an Old Oxonian . . . London, Belfast, William Mullan, 1880. [iv], 90 pp. Lettered on upper cover: Dick's / Biography / Isabel Burton.

356 k

Saint-Simon, L. de R. Duc de. Mémoires de Monsieur le Duc de S. Simon: ou, l'observateur véridique, sur le règne de Louis XIV . . . A Londres, Paris, Buisson, [etc.], 1789. 3 vols.

357

— Supplément aux Mémoires. Londres, Paris, Buisson, [etc.], 1789. 3 vols.

358

Sakcinski, I. K. Andreas Medulić Schiavone, Maler und Kupferstecher. Agram, Carl Albrecht, printer, 1863. [iii], 24 pp., front. (port.). [*see* p.].

359

— Borba Hrvatah u tridesetoljetnom ratu; spisao. Zagreb, Tiskom Dragutina Albrechta, 1874. [ii], 47 pp., front. (port.). Reprinted from *Arkiva* [*see* p.].

360

— Leben des G. Julius Clovio: ein Beitrag zur slawischen Kunstgeschichte; aus dem Ilirischen übersetzt von M. P. Agram, Franz Suppan, 1852. xii, 76 pp., front. (port.). [*see* p.].

361-2

Società Zoofila Triestina. Bollettino. 31 dicembre 1877, Anno XV, Punt. IV, pp. 49–64; and Anno XVII, Punt. III/IV, 15 dicembre 1879, pp. 25–40. On IB. [*see* 219].

363

Soyer, A. Memoirs of Alexis Soyer; with unpublished receipts and odds and ends of gastronomy; compiled and edited by F. Volant & J. R. Warren. London, W. Kent, 1859. xvi, 303 pp.

364

Stisted, G. M. Reminiscences of Sir Richard Burton. [London, 1891]. 335–42 pp. Reprinted from *Temple Bar,* Vol. 92.

365 k

Sully, M. de B., Duc de. Memoirs of Maximilian de Bethune, Duke of Sully, Prime Minister to Henry the Great . . . translated from the French . . . fourth edition. London, A. Millar, [etc.], 1763, 6 vols.

366

Times of Morocco, The. Tangier. Saturday, 13 Feb. 1886, No. 15. Reference to the Burtons on p. [3]. [*see* p.].

367

Ward, Mrs. Facts connected with the last hours of Napoleon. [?London, *c.* 1861]. 9 pp. [*see* p.].
Two other copies. [*see* 222, 238].

368 l

Ward, C. A. Oracles of Nostradamus. London, Leadenhall Press, [etc.]; New York, Scribner & Welford, [1891]. 375 pp.

369

[Ware, J. R.]. Wonderful dreams of remarkable men and women. London, Diprose & Bateman, [1883]. 96 pp. [*see* p.].

370

Whitfield, G. C. ed. Men of mark: a gallery of contemporary portraits. Vol. 1, No. II. London, Sampson Low, Marston, [etc.], 1876. [2] leaves, plates. [*see* pq.].

371

Wiseman, N. P. S. Recollections of the last four Popes, and of Rome in their times. London, Hurst & Blackett, 1858. x, 532 pp., front. (port.), plates.

372–448 Geography, Travel

372
Anonymous. L'art de voyager utilement suivant la copie de Paris. Amsterdam, J. Louis de Lorme, 1698. 52 pp. [*see* 1842].

373 a d h
Academia Real das Sciencias. Collecção de noticias para a historia e geografia das nações ultramarinas, que vivem nos dominios portuguezes, ou lhes São Visinhas. Lisboa, Typografia da Mesma Academia, 1812–56. 7 vols.

374
Back, Sir G., R. Collinson and Sir F. Galton, eds. Hints to travellers; third and revised edition; edited by a Committee of Council of the Royal Geographical Society. London, William Clowes, printer, 1871. 78 pp., illus. [*see* p.].

375 i
Baddeley, W. St C. Travel-tide. London, Sampson Low, Marston, [etc.], 1889. xi, 270 pp.

376 l
Bent, S. An address delivered before the St. Louis Historical Society . . . 1868, and . . . the Mercantile Library Association . . . 1869 upon thermometric gateways to the pole. Saint Louis, R. P. Studley, printer, 1869. 29 pp., map. [*see* p.].

377 d l
[Churchill, A. and J. Churchill]. A collection of voyages and travels, some now first printed from original manuscripts, others translated out of foreign languages, and now first published in English. London, Awnsham & John Churchill, 1704–32. 6 vols. illus., plates, maps. Vol. 3 dated 1703; wanting Vol. 5. [*see* q.].

378 d
Cook, T. Christmas supplement to Cook's excursionist and tourist advertiser, for the season of 1869. [London], 1869. 48 pp., map. [*see* 228].

379 d
Cooley, W. D. Dr. Livingstone and the Royal Geographical Society. London, printed for the author, 1874. 73 pp. [*see* p.].

380 a
Dionysius, *Periegetes*. Dionyssii Aphri de totius orbis situ, Antonio Becharia Veronensi interprete, consumatissimum opus. Jonnis Praeterea Honteri coronensis de cosmographiae rudimentis libri duo. Coelorum partes, stellas cum flatibus, amnes, regnaque cum populis, parue libelle tenes. [Basileae], excudebat Henricus Petrus Basileae, 1533. vii, 100 pp.

381 d f i
Galton, Sir F. The art of travel: or, shifts and contrivances available in wild countries; second edition. London, John Murray, 1856. xi, 248 pp.

382 i
— — third edition, revised and enlarged. London, John Murray, 1860. xviii, 298 pp., illus.

383 d
— ed. Vacation tourists and notes of travel in 1861. Cambridge, London, Macmillan, 1862. viii, 418 pp., illus., maps.

384 f k
Guthrie, W. A new geographical, historical and commercial grammar . . . third edition. London, printed for J. Knox, 1771. 728 pp., front. (maps), plates, illus., maps.
Hakluyt Society. Works issued by:

385 a d

Vol. 22. Major, R. H. ed. India in the fifteenth century: being a collection of narratives of voyages to India, in the century preceding the Portuguese discovery of the Cape of Good Hope . . . edited, with an introduction. London, 1857. xc, [140] pp.

386 a d

Vol. 32. Barthema, L. The travels of Ludovico di Varthema in Egypt, Syria, Arabia deserta and Arabia felix, in Persia, India, and Ethiopia, A.D. 1503 to 1508; translated from the original Italian edition of 1510, with a preface by John Winter Jones, and edited, with notes and an introduction, by George Percy Badger. London, 1863. cxxvi, 320 pp., front. (map), illus., map.

387 a d (Vol. 1)

Vols. 33, 68. Cieza de Leon, P. de. The travels of . . . A.D. 1532–50, contained in the first part of his Chronicle of Peru; translated and edited by Clements R. Markham. London, 1864–83. 2 vols., front. (map).

Title of Vol. 2: The second part of the Chronicle of Peru.

388 a d

Vol. 34. Andagoya, P. de. Narrative of the proceedings of Pedrarias Davila in the provinces of Tierra Firme or Castilla del Oro, and of the discovery of the South sea and the coasts of Peru and Nicaragua; translated and edited . . . by Clements R. Markham. London, 1865. xxix, 88 pp.

389 a d

Vol. 35. Barbosa, D. A description of the coasts of east Africa and Malabar in the beginning of the sixteenth century; translated . . . with notes and preface, by Henry E. J. Stanley. London, 1866. xi, 336 pp., plates.

390 d

Vol. 38. Collinson, Sir R. The three voyages of Martin Frobisher, in search of a passage to Cathaia and India by the North-west, A.D. 1576–8; reprinted from the first edition of Hakluyt's Voyages, with selections from manuscript documents in the British Museum and State Paper Office. London, 1867. xxvi, 376 pp., front. (port.), maps.

391 d

Vol. 39. Morga, A. de. The Philippine islands, Moluccas, Siam, Cambodia, Japan, and China, at the close of the sixteenth century; translated from the Spanish, with notes and a preface, and a letter from Luis Vaez de Torres, describing his voyage through the Torres straits, by Henry E. J. Stanley. London, 1868. xxiv, 431 pp., front. (port.), illus., plate.

392 d

Vol. 40. Cortes, H. The fifth letter . . . to the Emperor Charles V, containing an account of his expedition to Honduras; translated from the original Spanish by Don Pascual de Gayangos. London, 1868. xvi, 156 pp.

393 a d

Vols. 41, 45. La Vega, G. Lasso de, El Inca. First part of the Royal commentaries of the Yncas by the Ynca Garcilasso de la Vega; translated and edited . . . by Clements R. Markham. London, 1869–71. 2 vols. front.

394 a d h

Vol. 42. Correa, G. The three voyages of Vasco de Gama, and his viceroyalty from the Lendas da India . . . translated from the Portuguese, with notes and an introduction, by Henry E. J. Stanley. London, 1869. lxxx, 430, xxxvi pp., front. (port.), plates.

395 a d

Vol. 43. Columbus, C. Select letters of Christopher Columbus, with other original documents relating to his four voyages to the New world; translated and edited by R. H. Major; second edition. London, 1870. cxlii, 254 pp. front., maps, table, bibliogr.

396 d h

Vol. 44. Salîl ibn Ruzaik. History of the Imâms and Seyyids of 'Omân, by Salîl-ibn-Razîk, from A.D. 661–1856; translated from the original Arabic, and edited, with notes . . . and an introduction, continuing the history down to 1870, by George Percy Badger. London, 1871. cxxviii, 436 pp., front. (map), bibliogr.

Vol. 45 *see* 393.

397 a d

Vol. 46. Boutier, P. and J. Le Verrier. The Canarian, or book of the conquest and conversion of the Canarians in the year 1402, by Jean de Bethencourt . . . composed by Pierre Bontier [sic] and Jean Le Verrier . . . translated and edited, with notes and an introduction, by Richard Henry Major. London, 1872. lv, 229 pp., front. (port.), plates, map.

398 a d f
Vol. 47. Markham, Sir C. R. ed. Reports on the discovery of Peru . . . translated and edited, with notes and an introduction . . . London, 1872. xxiii, 143 pp., front. (map).

399 a
Vol. 48. Markham, Sir C. R. ed. Narratives of the rites and laws of the Yncas; translated from the original Spanish manuscripts, and edited, with notes and an introduction. London, 1873. xx, 220 pp., illus., plate, map.

400
Vol. 49 [1]. Barbaro, J. and A. Contarini. Travels to Tana and Persia; translated from the Italian by William Thomas . . . and by S. A. Roy . . . edited with an introduction by Lord Stanley of Alderley. London, 1873. xi, 175 pp.

401
Vol. 49 [2]. Grey, C. ed. A narrative of Italian travels in Persia in the fifteenth and sixteenth centuries; translated and edited. London, 1873. xvii, 231 pp. [*see* 400].

402 d
Vol. 50. Zeno, N., the elder and A. Zeno. The voyages of the Venetian brothers Nicolò & Antonio Zeno, to the northern seas, in the XIVth century, comprising the latest known accounts of The lost colony of Greenland [by I. Bardsen]; and of the Northmen in America before Columbus; translated and edited, with notes and an introduction by Richard Henry Major. London, 1873. cv, 64 pp., plate, maps, table.

Vol. 51 *see* 44.

403 a d
Vols. 53, 55, 62, 69. Albuquerque, A. de', the younger. The commentaries of the great Afonso Dalboquerque, second Viceroy of India; translated from the Portuguese edition of 1774, with notes and an introduction, by Walter de Gray Birch. London, 1875–84. 4 vols. fronts., illus., plates, maps.

404 h
Vol. 54. Veer, G. de. The three voyages of William Barents to the Arctic regions (1594, 1595, and 1596); second edition, with an introduction by Koolemans Beynen. London, 1876. clxxvi, 289 pp., front. (map), plates, maps.

Vol. 55 *see* 403.

405 d
Vol. 56. Markham, Sir C. R. ed. The voyages of Sir James Lancaster to the East Indies, with abstracts of journals of voyages to the East Indies, during the seventeenth century, preserved in the India Office; and the voyage of Captain John Knight (1606) to seek the North-west passage. London, 1877. xxii, 314 pp.

406
Vol. 57. Markham, Sir C. R. ed. The Hawkins' voyages during the reigns of Henry VIII, Queen Elizabeth and James I; edited with an introduction. London, 1878. lii, 453 pp., front. (port.), illus.

407 a d
Vol. 59. Davis, J. The voyages and works of John Davis, the Navigator; edited with an introduction and notes by Albert Hastings Markham. London, 1880. xcv, 392 pp. front., illus., plate, maps.

408
Vols. 60–1. Acosta, J. de. The natural & moral history of the Indies; reprinted from the English translated edition of Edward Grimston, 1604, and edited, with notes and an introduction, by Clements R. Markham. London, 1880. 2 vols., maps.

Vol. 62 *see* 403.

409 a d
Vol. 63. Baffin, W. The voyages of William Baffin, 1612–1622; edited, with notes and an introduction, by Clements R. Markham. London, 1881. lix, 192 pp., front. (port.), maps, tables.

410 a d
Vol. 64. Alvares, F. Narrative of the Portuguese embassy to Abyssinia during the years 1520–1527; translated from the Portuguese, and edited with notes and an introduction, by Lord Stanley of Alderley. London, 1881. xxviii, 416 pp., map.

411 a d
Vol. 65. Lefroy, Sir J. H. ed. The historye of the Bermudaes, or Summer islands; edited from a MS. in the Sloane collection, British Museum. London, 1882. xii, 327 pp., front. (port.), plates.

412 a d

Vols. 66–7. Cocks, R. Diary of Richard Cocks, Cape-merchant in the English factory in Japan, 1615–1622; with correspondence edited by Edward Maunde Thompson. London, 1883. 2 vols.

Vol. 68 *see* 387.

Vol. 69 *see* 403.

413 a d

Vols. 70–1. Linschoten, J. H. van. The voyage of . . . to the East Indies, from the old English translation of 1598; the first book containing his description of the East, edited . . . by the late Arthur Coke Burnell . . . the second volume, by P. A. Tiele. London, 1885. 2 vols. front. (port.).

414

Vols. 72–3. Jenkinson, A. and others. Early voyages and travels to Russia and Persia . . . with some account of the first intercourse of the English with Russia and central Asia by way of the Caspian sea; edited by E. Delmar Morgan. London, 1886. 2 vols. front. (port.), illus., maps.

415 a d

Vols. 74–5, 78. Hedges, Sir W. The diary of William Hedges . . . during his Agency in Bengal, as well as on his voyage out and return overland (1681–1687); transcribed . . . with introductory notes, etc., by R. Barlow, and illustrated by copious extracts from unpublished records, etc., by Henry Yule. London, 1887–9. 3 vols. fronts., plates.

416 a d

Vols. 76–7. Pyrard, F. The voyage of François Pyrard of Laval to the East Indies, the Maldives, the Moluccas and Brazil; translated into English from the third French edition of 1619, and edited with notes, by Albert Gray, assisted by H. C. P. Bell. London, 1887–8. 2 vols. fronts., illus., plates.

Vol. 78 *see* 415.

417 h

Vols. 82–3. Oliver, S. P. ed. The voyage of François Leguat of Bresse to Rodriguez, Mauritius, Java, and the Cape of Good Hope; transcribed from the first English edition; edited and annotated by Pasfield Oliver. London, 1891. 2 vols. fronts., illus., plates, maps.

418

Vol. 86. Columbus, C. The journal of Christopher Columbus (during his first voyage, 1492–93), and documents relating to the voyages of John Cabot and Gaspar Corte Real; translated, with notes and an introduction, by Clements R. Markham. London, 1893. liv, 259 pp., plate, maps.

419

Vol. 87. Bent, J. T. ed. Early voyages and travels in the Levant. I, The diary of Master Thomas Dallam, 1599–1600 – II, Extracts from the diaries of Dr. John Covel, 1670–1679 . . . edited, with an introduction and notes. London, 1893. xlv, 305 pp., front. (port.).

420

Vols. 88–9. Foxe, L. and T. James. The voyages of Captain Luke Foxe of Hull, and Captain Thomas James of Bristol, in search of a North-west passage, in 1631–32 . . . edited, with notes and an introduction by Miller Christy . . . London, 1894. 2 vols. fronts. (ports.), illus., plate, maps.

421

Vol. 90. Vespucci, A. The letters of Amerigo Vespucci and other documents illustrative of his career; translated, with notes and an introduction, by Clements R. Markham. London, 1894. xliv, 121 pp., illus.

422

Vol. 91. Sarmiento de Gambóa, P. Narratives of the voyages . . . to the Straits of Magellan; translated and edited, with notes and an introduction by Clements R. Markham. London, 1895. xxx, 401 pp., maps.

423

Vols. 91–4. Leo, Africanus. The history and description of Africa and of the notable things therein contained, written by Al-Hassan ibn-Mohammed al-Wezaz al-Fasi, a Moor . . . known as Leo Africanus; done into English in the year 1600, by John Pory, and now edited, with an introduction and notes by Robert Brown. London, 1896. 3 vols. maps.

424 a b d

Herschel, Sir J. F. W. Physical geography from the Encyclopaedia britannica. Edinburgh, Adam & Charles Black, 1861. viii, 441 pp., illus., maps.

425

Ibrāhīm ibn Yahya, al-Nakkāsh, called Ibn al-Zarkālah. Sapheae recentio res doctrinae patris Abrusahk Azarchelis summi astronomi. [Norimberge, excusum, 1534]. [26] pp., tables. [*see* 380].

426 d f h

Jackson, J. R. What to observe: or, the traveller's remembrancer; second edition. London, Madden & Malcolm, 1845. xix, 570 pp., illus. [*see* 1360].

427 d

— — third edition, revised and edited by Norton Shaw. London, Houlston & Wright, 1861. xii, 538 pp., illus.

428

Johnson, A. K. A general dictionary of geography, descriptive, physical, statistical, historical, forming a complete gazetteer of the world; new edition. London, Longmans, Green, 1877. [vii], 1513 pp., tables.

429 b

Maury, M. F. The physical geography of the sea; new edition . . . London, Sampson Low, 1859. xxiv, 352 pp., plates, maps, tables.

430 a d

Mela, P. Geografia di Pomponia Mela; libri tre, tradotti ed illustrati da Giovanni Francesco Muratori. Torino, Stamperia Reale, 1855. xlviii, 267 pp.

431

— Pomponii Melae de situ orbis; libri tres, cum indice. [Norimbergae, 1556]. [112] pp. [*see* 380].

432

Milne, J. Across Europe and Asia: travelling notes. [Hertford, Stephen Austin, printer], 1877–8. 88 pp., illus. Reprinted from the *Geological magazine,* Decade 2, Vols. 4–5. [*see* p.].

433 a d

Mueller, C. Geographi graeci minores; volumen primum. Parisiis, Ambrosio Firmin Didot, 1855. cxlv, 577, vii pp., maps.

434 a e

Negri, C. Discorso del . . . Società geografica italiana . . . 13 marzo 1870. [Rome], 1870. 72 pp. [*see* 257].

435

— — 30 aprile 1871. [Rome], 1871. lv pp. [*see* 257].

436 i

Nolte, V. Fifty years in both hemispheres: or, reminiscences of the life of a former merchant; translated from the German. New York, Redfield, 1854. xxii, 476 pp. (Charles H. Eastman, New Orleans).

437 a d e

Pinkerton, J. A general collection of the best and most interesting voyages and travels in all parts of the world, many of which are now first translated into English . . . London, Longman, Hurst, [etc.], 1808–14. 17 vols. fronts., plates, maps, tables.

438

— Modern geography: a description of the empires, kingdoms, states, and colonies . . . in all parts of the world . . . a new edition. London, T. Cadell & W. Davies, [etc.], 1817. 2 vols.

439 a d

Ptolemaeus, C. Geographia; edidit Carolus Fridericus Augustus Nobbe . . . editio stereotypa. Lipsiae, Sumptibus et Typis Caroli Tauchnitii, 1843–5. 3 vols. (in 1). illus., map.

440

Roesler, R. Die Aralseefrage; noch einmal geprüft. Wien, Karl Gerold's Sohn, 1873. [ii], 88 pp. [*see* 230].

441 i

[Salvator, A. L.]. Um die Welt ohne zu Wollen . . . Würzburg, Wien, Leo Woerl, 1883. viii, 343 pp., plates.

442 k

Seton-Karr, H. W. Ten years travel & sport in foreign lands: or, travels in the eighties. London, Chapman & Hall, 1890. xi, 445 pp., front. (port.).

443 k

[Smollett, T. G. compiler]. Compendium of authentic and entertaining voyages, digested in a chronological series, the whole exhibiting a clear view of the customs, manners, religion, government . . . of most nations in the known world. London, R. & J. Dodsley, [etc.], [1756]. 7 vols. front., plates, maps.

444 d l

Societa geografica italiana. Elenco delle questioni presentate al III° Congresso geografico internazionale. Roma, Giuseppe Civelli, 1881. 50 pp. [*see* p.].

445

Société de Geographie. 1861 – XLIe année: listes des membres. Paris, L. Martinet, 1861. 32 pp. [*see* 238].

446

Stevens, S. Directions for collecting and preserving specimens of natural history, in tropical climates; second edition. [London, M'Gowan, 187–]. 8 pp., illus. [*see* 228].

447 a d

Strabo. The geography of Strabo, literally translated, with notes . . . by H. C. Hamilton . . . W. Falconer. London, Henry G. Bohn, 1854–7. 3 vols. (Bohn's classical library).

448 d

Suez. Guida provvisoria pei viaggiatori in Brindisi in occasione dell'apertura del canale di Suez. [Brindisi, Tip. adriatico-orientale, 1869]. 31 pp., tables. [*see* 257].

449-660 Linguistics

General

449 a d g
Ballhorn, R. Grammatography: a manual of reference to the alphabets of ancient and modern languages based on the German compilation of F. Ballhorn. London, Trübner, 1861. 76 pp.

450 k
Bell, D. C. and A. M. Bell. Bell's Standard elocutionist . . . followed by . . . extracts in prose and poetry; new edition. London, William Mullan, 1878. 510 pp. (Mullan's standard educational works).

451 a d
Bopp, F. A comparative grammar of the Sanskrit, Zend, Greek, Latin, Lithuanian, Gothic, German, and Sclavonic languages; translated from the German by Edward B. Eastwick; third edition. London, Edinburgh, Williams & Norgate, 1862. 3 vols.

452 d i
Charnock, R. S. Local etymology: a derivative dictionary of geographical names. London, Houlston & Wright, 1859. x, 325 pp. (24 Dec. 1864).

453 i
— Verba nominalia: or, words derived from proper names. London, Trübner, 1866. iv, 357 pp.

454 a d h
Iwanowitsch, I. Die Weltsprache Volapük in drei Lectionen. Leipzig, Eduard Heinrich Mayer, [1887]. 28 pp. [*see* p.].

455
Obermueller, W. Deutsch-keltisches, geschichtlich-geographisches Wörterbuch zur Erklaerung der Fluss- Berg- Orts- Gau – Völker- und Personen-Namen Europas, West-Asiens und Nord-Afrikas im Allgemeinen wie insbesondere Deutschlands nebst den daraus sich ergebenden Folgerungen für die Urgeschichte der Menscheit. Berlin, Deinick's Verlag Link & Reinke, [etc.], 1872. 2 vols.

456 d i
Reinisch L. Der einheitliche Ursprung der Sprachen der alten Welt nachgewisen durch Vergleichung der afrikanischen, erythräischen und indogermanischen Sprachen mit zugrundelegung des Teda. Erster Band. Wien, Wilhelm Braumüller, 1873. xviii, 408 pp., illus.

457 a d
Sayce, A. H. The principles of comparative philology; third edition, revised and enlarged. London, Trübner, 1885. xlviii, 422 pp.

458
Schleyer, J. M. Mittlere Grammátik der Universàlspràche Volapük; achte Auflage. Konstanz in Baden, Schleyer's Weltspràche-Zentràlbüro, 1887. viii, 144 pp. At head of title: Wéltspràche. [*see* p.].

459 a d h
Schneid, A. Vereinfachtes Volapük: praktischer Leitfaden für den Selbstunterricht. Brünn, C. Winkler, 1887. 56 pp. [*see* p.].

460
Truebner & Co. A catalogue of dictionaries and grammars of the principal languages and dialects of the world for sale . . . London, Trübner, 1872. 64, iii pp. [*see* 257].

461 a d i
Twisleton, E. T. B. The tongue not essential to speech; with illustrations of the power of speech in the African confessors. London, John Murray, 1873. iv, 332 pp.

462 a

Beltrame, G. Grammatica della lingua denka. Firenze, Stabilimento G. Civelli, 1870. 159 pp.

463 d h i

Bowen, T. J. Grammar and dictionary of the Yoruba language; with an introductory description of the country and people of Yoruba. Washington, Smithsonian Institution, 1858. xxiii, 71, 136 pp., map, tables. (Smithsonian contributions to knowledge).

464 d

Bunsen, G. De Azania Áfricae littore orientali: commentatio philologica. Bonnae, Formis Caroli Georgii, 1852. iii, 44 pp., map. [*see* 237].

465 a d

Cannecattim, B. M. de. Diccionario da lingua bunda, or angolense, explicada na portuguesa, e latina. Lisboa, Impressão Regia, 1804. [5], ix, 722 pp.

466 a d

Christaller, J. G. A dictionary of the Asante and Fante language, called Tshi (Chwee, Tw̌i); with a grammatical introduction and appendices on the geography of the Gold Coast. Basel, Evangelical Missionary Society, 1881. xxviii, 671 pp.

467 d

Clarke, H. Memoir on the comparative grammar of Egyptian, Coptic, & Ude. London, Trübner, 1873. 31 pp. [*see* p.].

468 d

Clarke, J. Introduction to the Fernandian tongue. Part 1; second edition. Berwick-on-Tweed, Daniel Cameron, printer, 1848. 56 pp.

469

Committee of Friends for Promoting African Instruction. African lessons: Wolof and English, in three parts. Part first: easy lessons, and narratives for schools. London, William Phillips, 1823. [xvii], 55 pp. [*see* 239].

470 a d

Crowther, S. A grammar and vocabulary of the Yoruba language . . . together with introductory remarks by O. E. Vidal. London, Seeleys, 1852. v, 38, vii, 52, [ii], 291 pp.

471 a d

— A vocabulary of the Yoruba language . . . together with introductory remarks by O. E. Vidal. London, Seeleys, 1852. v, 38, [ii], 291 pp.

472

Dard, J. Grammaire wolofe; ou méthode pour étudier la langue des noirs qui habitent les royaumes de Bourba-Yolof, de Walo . . . [Paris], Imprimerie Royale, 1826. [4], xxxi, 213 pp., map.

473 a d

Doehne, J. L. A Zulu-Kafir dictionary etymologically explained, with copious illustrations and examples; preceded by an introduction on the Zulu-Kafir language. Cape Town, G. J. Pike's Machine Printing Office, 1857. xlii, 417 pp.

474

Erhardt, J. Vocabulary of the Enguduk Iloigob, as spoken by the Masai-tribes in East-Africa. Ludwigsburg in Würtemberg, Ferdinand Riehm, printer, 1857. 111 pp.

475

Freeman, H. S. A grammatical sketch of the Temahuq or Towarek language. London, Harrison, 1862. 47 pp.

476 a d

Gaboon Mission. A grammar of the Mpongwe language, with vocabularies, by the Missionaries of the A.B.C.F.M. Gaboon Mission, Western Africa. New York, Snowden & Prall, 1847. 94 pp., tables.

477

Girard, A. Glassaire abyssin. Marseille, Marius Olive, printer, 1877. 11 pp. Reprinted from the *Compte-rendu du Congrès des Orientalistes de Marseille, 2e Session des Congrès provinciaux des Orientalistes*, 1876. [*see* p.].

478 a d

Goldie, H. Dictionary of the Efïk language, in two parts. I, Efïk and English; II, English and Efïk. Glasgow, Dunn & Wright, printer, 1862. li, 643 pp.

479 a d

— Principles of Efïk grammar, with specimen of the language. Old Calabar, Mission Press, printer, 1857. xx, 86 pp.

480
Grey, Sir G. The library of His Excellency Sir George Grey: philology. Vol. 1. London, Trübner; Leipzig, F. A. Brockhaus, 1858. 2 vols. (in 1).

481
Grout, L. The Isizulu: a grammar of the Zulu language; accompanied with a historical introduction, also with an appendix. Pietermaritzburg, May & Davis, [etc.]; London, Trübner, 1859. lii, 432 pp., illus.

482 h
Hanoteau, A. Essai de grammaire kabyle, renfermant les principes du langage parlé par les populations du versant nord du Jurjura et spécialement par les Igaouaouen ou Zouaoua. Alger, Bastide, [etc.]; Paris, Challamel, Benjamin Duprat, [1858]. xxiv, 395 pp., tables.

483 a d
Jaubert, P. A. E. P. Grammaire et dictionnaire abrégés de la langue berbère; composés par feu Venture de Paradis . . . Paris, Imprimerie Royale, 1844. xxiii, 236 pp. (Recueil de voyages et de mémoires, publié par la Société de Géographie, Vol. 7, no. 1).

484 d
Krapf, J. L. Outline of the elements of the Kisuáheli language, with special reference to the Kiníka dialect. Tübingen, Lud. Fried. Fues, printer, 1850. 142 pp.

485 a e
— Vocabulary of six East-African languages (Kisuáheli, Kiníka, Kikámba, Kipokómo, Kihiáu, Kigálla. Tübingen, Ludw. Friedr. Fues, 1850. x, 64 pp. [*see* q.].

486
— and J. Rebmann. The beginning of a spelling book of the Kinika language; accompanied by a translation of the Heidelberg catechism. Bombay, American Mission Press, 1848. 78 pp. [*see* 232].

487 i
Le Berre, —. Grammaire de la langue pongouée. Paris, Imprimerie Simon Raçon, 1873. [3], iv, 223 pp. (B. B. N. Walker).

488 a d
Macbriar, R. M. Grammar of the Fulah language; from a MS.. . . in the British Museum; edited, with additions, by E. Norris. London, Harrison, 1854. viii, 95 pp. Proof copy.

489 a d l
— A grammar of the Mandingo language; with vocabularies. London, Wesleyan-Methodist Missionary Society, [1842?]. viii, 74 pp.

490 a d
Macdonald, W. B. Sketch of a Coptic grammar adapted for self-tuition. Edinburgh, W. H. Lizars; London, George Philip, 1856. [2], ii, 54 pp., table.

491
Mackey, J. L. A grammar of the Benga language. New York, Mission House, [1855]. [ii], 60 pp., front. [*see* 231].

492
[Merrick, J.]. A dictionary of the Isubu tongue. [Bimbia, 1842]. [ii, 432] pp.

493
[Newman, F. W.]. A grammar of the Berber language. [Bonn], 1845. 245–336 pp. Reprinted from the *Zeitschrift für die Kunde des Morgenlandes,* Vol. 6, no. 2. [*see* 239].

494
[Norris, E.]. Grammar of the Bornu or Kanuri language; with dialogues, translations, and vocabulary. London, Harrison, printer, 1853. 101 pp. Pp. 47–74 entitled: Grammatical sketch of the Bornu or Kanuri language. [*see* p.].

495 a d
Payne, J. A dictionary of the Grebo language. New York, Edward O. Jenkins, printer, 1860. 100 pp.
Another copy.

496
— Grebo primer, for the use of the Protestant Episcopal Mission. [New York], American Tract Society, 1860. 70 pp., illus.

497 a d
[Preston, — and — Best]. A grammar of the Bakĕle language, with vocabularies, by the Missionaries of the A.B.C.F.M., Gaboon Station, Western Africa. New York, J. P. Prall, printer, 1854. 117 pp. imperfect.

498
Reinisch, L. Aegyptische Chrestomanthie . . . in drei Lieferungen. 1. Lieferung

[prospectus]. Wien, Wilhelm Braumüller, 1873. [i] p., 21 plates. [*see* q.].

499 d

— Die Nuba-Sprache. Wien, Wilhelm Braumüller, 1879, 2 vols. (in 1). Contents: 1, Grammatik und Texte – 2, Nubisch-deutsches und deutsch-nubisches Wörterbuch.

500 a d

Riis, H. N. Grammatical outline and vocabulary of the Oji-language, with especial reference to the Akwapim-dialect, together with a collection of proverbs of the natives. Basel, Bahnmaier's Buchhandlung (C. Detloff), 1854. viii, 276 pp. [*see* 506].

501 i

[Saker, A. J. S.]. A vocabulary of the Dualla language, for the use of missionaries and others. Cameroons, Mission Press, 1862. [i], 63 pp. [*see* 239].

502

— ed. A grammar of the Isubu tongue. [n.p., 1852]. 41 pp., table. [*see* 238].

503 a d

Schoen, J. F. Grammar of the Hausa language. London, Church Missionary House, 1862. [2], xiv, viii, 234 pp.

504 a d

— Oku Ibo: grammatical elements of the Ibo language. London, [W. M. Watts], 1861. [ii], 8, 86 pp.

505

— Vocabulary of the Haussa language. Part I, English and Haussa; part II, Haussa and English; and phrases, and specimens of translations to which are prefixed, the grammatical elements of the Haussa language. London, Church Missionary Society, 1843. [ix], 190 pp.

l

Another copy.

505.1

Schreuder, H. P. S. Grammatik for Zulu-sproget . . . med fortale og anmaerkninger af C. A. Holmboe. Christiania, W. C. Fabritius, 1850. viii, 88 pp. [*see* 239].

506 a d

Zimmermann, J. A grammatical sketch of the Akra- or Gã-language, with some specimens of it from the mouth of the natives and a vocabulary of the same, with an appendix on the Adaṅme-dialect. Stuttgart, Basel Missionary Society; J. F. Steinkopf, printer, 1858. 2 vols. (in 1), tables.

Linguistics – *America*

507 g

Febres, A. Gramatica chilena. Concepcion, Imprenta de la Union, 1864. [3], iii, 77 pp.

508 e i

Ferreira França, E. Chrestomathia da lingua brazilica. Leipzig, F. A. Brockhaus, Liveiro de S.M.O. Imperador do Brazil, 1859. xviii, 230 pp. (Bibliotheca brasilienze, Vol. 3; Bibliotheca linguistica, Vol. 2). (M. Prado, S[ão] P[aulo], 18 Febr. 1866).

509 a

Figueira, L. Arte da grammatica da lingua do Brasil; quarta impressaõ. Lisboa, na Officina Patriarcal, 1795. [iii], 101 pp.

510 a d l

Gonçalves Dias, A. Diccionario da lingua tupy chamada, lingua geral dos indigenas do Brazil. Lipsia, F. A. Brockhaus, Livreiro de S.M.O Imperador do Brazil, 1858. viii, 191 pp.

511 a d

Ludewig, H. E. The literature of American aboriginal languages; with additions and corrections by Wm. W. Turner; edited by Nicolas Trübner. London, Trübner, 1858. xxiv, 258 pp. (Trübner's Bibliotheca glottica, Vol. 1).

512 a d

Markham, Sir C. R. Contributions towards a grammar and dictionary of Quichua; the language of the Yncas of Peru. London, Trübner, 1864. v, 223 pp.

513 a d

[Pereira Coruja, A. A. intro.]. Collecção de vocabulos e frases usados na Provincia de S. Pedro do Rio Grande do Sul no Brazil. Londres, Trübner, 1856. 32 pp. [*see* 253]. Another copy. [*see* p.].

514 d i

Silva Guimarães, J. J. Grammatica da lingua geral dos Indios do Brasil; reimpressa pela primeira vez neste continente depois de tão longo tempo de sua publicação em Lisboa. Bahia, Typographia de Manoel Feliciano Sepulveda, 1851. [12], vi, 105, 15 pp. (M. Prado).

Linguistics – *Asia*

515
Anonymous. Nor aybbenaran ev hegerēn: hayerēn šitak hegel sorvelu, dpratanc tloc hamar [A new alphabet primer and spelling book: to learn to spell Armenian correctly, for the boys of the school; Armenian text. Venice, Armenian Monastery of S. Lazar, printer], 1850. [xvi], 80 pp. [*see* 260].

516 a d
'Abd Allah ibn Muhammad, al-Shubrāwī. Risālah latīfah fi qawā'id al-nahw [A letter on the rules of grammar; Arabic text]. Damascus, Kanīsat Irland al-qusūsīyah, [Irish Protestant Church Press], 1864. 12 pp. [*see* p.].

517
Ayrton, F. collector. A descriptive catalogue of the Oriental caligraphs, &c., . . . compiled in 1874, by Máz-Haru'd-dîn Ásaad . . . translated and arranged by George Percy Badger. London, William Whitely, printer, 1885. v, 98 pp.

518 i
Badger, G. P. An English-Arabic lexicon, in which the equivalents for English words and idiomatic sentences are rendered into literary and colloquial Arabic. London, C. Kegan Paul, 1881. xii, 1248 pp. (5 Oct. 1885). [*see* q.].

519 a d
Barker, W. B. A practical grammar of the Turkish language; with dialogues and vocabulary. London, Bernard Quaritch, 1854. [vi], 160 pp.

520 a d (p. 36)
— A reading book of the Turkish language, with a grammar and vocabulary. London, James Madden, 1854. xxiv, 101, [106], 56 pp., tables.

521 a d
Bleeck, A. H. A concise grammar of the Persian language, containing dialogues, reading lessons, and a vocabulary; together with a new plan for facilitating the study of languages. London, Bernard Quaritch, 1857. 2 pts (in 1).

522 a d l
Caldwell, R. A comparative grammar of the Dravidian or South-Indian family of languages; second edition. London, Trübner, 1875. xli, 154, 608 pp., tables.

523 a l
Catafago, J. An English and Arabic dictionary in two parts: Arabic and English, and English and Arabic; second edition. London, Bernard Quaritch, 1873. viii, 1096 pp., tables.

524 a d
Caussin de Perceval, A. P. Grammaire arabe vulgaire, pour les dialectes d'Orient et de Barbarie. Paris, Librairie orientale de Dondey-Dupré, 1833. xvi, 172, [10] pp.

525 a d
Chamberlain, B. H. A simplified grammar of the Japanese language; modern written style. London, Trübner, [etc.], 1886. viii, 105 pp. (Trübner's collection of simplified grammars, Vol. 15).

526 d
Cowper, B. H. The principles of Syriac grammar; translated and abridged from the work of Dr. Hoffmann. London, Williams & Norgate, [etc.], 1858. xvi, 184 pp., tables.

527 a d
[Davidson, B.]. Syriac reading lessons; consisting of copious extracts from the Peschito version of the Old and New Testaments . . . grammatically analysed and translated with the elements of Syriac grammar . . . London, Samuel Bagster, [1851]. [3], xxxvi 87 pp., tables.

528 a d
Edkins, J. A grammar of the Chinese colloquial language commonly called the Mandarin dialect; second edition. Shanghai, Presbyterian Mission Press, 1864. viii 279 pp., tables.

529 a d
— Introduction to the study of the Chinese characters. London, Trübner, 1876. xx 211, iv, 104 pp.

530 a d
Faris ibn Yūsuf, al-Shidyak. A practical grammar of the Arabic language, with inter

lineal reading lessons, dialogues and vocabulary. London, Bernard Quaritch, 1856. [v], 148 pp.

531 l

— — second edition, by the Rev. Henry G. Williams. London, Bernard Quaritch, 1866. [iv], 162 pp.

532

Forbes, D. Arabic reading lessons, consisting of easy extracts from the best authors . . . also some explanatory annotations, etc. London, Wm. H. Allen, 1864. viii, 53, [ciii], 76 pp.

533 d l

[Freytag, G. W. ed.]. Lexicon arabico-latinum ex opere sua majore . . . edidit G. W. F. Halae Saxonum, 1837. 694 pp. Title-page wanting.

534 a d

Gladwin, F. Dissertations on the rhetoric, prosody and rhyme of the Persians. Calcutta, London, Oriental Press for J. Debrett, 1801. [v], 171 pp.

535 d

Ḥasan, al-Kafrāwī. Kurzgefasste Grammatik der vulgär-arabischen Sprache, mit besonderer Rücksicht auf den egyptischen Dialekt, von A. Hassan. Wien, Druck und Verlag der K. K. Hof- und Staatsdruckerei, 1869. viii, 244, [20] pp.

536

— Sharḥ wa i'rāb al-Kafrāwī 'alā matn al-Ajurrūmīyah [Commentary on Ibn Ajurrūm's introduction to Arabic syntax entitled al-Muqaddimat al-Ajurrūmīyah; edited by Naṣr al-Hūrīnī; Arabic text]. Cairo, 1280 A.H. [1864 A.D.]. 216 pp.

537 a d

[Hindī-Sindhī Committee]. A writing book teaching the new "Improved Hindī-Sindhī alphabet", devised in 1868 by an official committee. Karachi, 1875. 38 pp. [*see* p.].

538 a j

Kremer, A. Beiträge zur arabischen Lexikographie. Wien, Carl Gerold, 1883. 92 pp. (Vienne, 12 Mai 1884). [*see* p.].

539 a d

Lerchundi, J. Rudimentos del árabe vulgar, que se habla en el Imperio de Marruecos. Madrid, Imprenta y Estereotipia de M. Rivadeneyra, 1872–3. 2 vols. (in 1).

540 a d

Lockett, A. The Miut Amil, and Shurhoo Mjut Amil: two elementary treatises on Arabic syntax; translated from the original Arabic; with annotations, philological and explanatory . . . Calcutta, P. Pereira at the Hindoostanee Press, printer, 1814. xxxiii, 235, [48] pp.

541 a d

Muhammad ibn 'Abd Allah (Jamal al-Dīn Abū 'Abd Allah), al Tai Jayyānī called Ibn Mālik. Alfiyya ou la quintessence de la grammaire arabe; ouvrage de Djémal-eddin Mohammed, connu sous le nom d'Ebn-Malec; publié en original, avec un commentaire, par . . . Silvestre de Sacy. Paris, London, Oriental Translation Fund of Great Britain & Ireland, 1833. viii, 254, [143] pp.

542 h

Neuphal, G. Dialogues ou guide de conversations en Arabe et en Français. Beyrouth, Imprimerie catholique, 1868. [ii], 528 pp.

543 d i

Palmer, E. H. Simplified grammar of Hindūstānī, Persian and Arabic. London, Trübner, 1882. vii, 104 pp. (Trübner's collection of simplified grammars, Vol. 1). (N. Trübner).

544 a d

Parkhurst, J. A Hebrew and English lexicon, without points; in which the Hebrew and Chaldee words of the Old Testament are explained in their leading and derived senses . . . to this work are prefixed a Hebrew and Chaldee grammar . . . a new edition. London, William Baynes, 1823. [2], xvi, 54, 677 pp., plate, table.

545 a d

Parmentier, T. De la transcription pratique au point de vue français des noms arabes en caractères latins. Paris, Secrétariat de l'Association [française pour l'Avancement des Sciences], 1880. [iii], 35 pp., tables. [*see* p.].

546 d i

— Vocabulaire arabe-français des principaux termes de géographie. Paris, Secrétariat de l'Association [française pour l'Avancement des Sciences], 1882. [iii], 50 pp. [*see* p.].

547

Redhouse, Sir J. W. An English and Turkish dictionary, in two parts, English and

Turkish, and Turkish and English. London, Bernard Quaritch, 1857. [v], 429–1149 pp.

548 l
— Grammaire raisonnée de la langue ottomane, suivie d'un appendice. Paris, Gide, 1846. [vi], 343 pp., tables.

549
— A lexicon, English and Turkish. London, Bernard Quaritch, 1861. [xix], 827 pp.

550 a d i l
Richardson, J. A dictionary, Persian, Arabic, and English; with a dissertation on the languages, literature, and manners of Eastern nations; revised and improved by Charles Wilkins; a new edition, by Francis Johnson. London, J. L. Cox, printer for Parbury, Allen, [etc.], 1829. 9, lxxxvi, 1714 pp. (Ali Akbar). [*see* q.].

551 a d g
— A grammar of the Arabic language . . . principally adapted for the service of the Honourable East India Company; a new edition. London, Lackington, Allen; Longman, Hurst, [etc.], 1811. xii, 212 pp.

552 a i
Salmoné, H. A. On the importance to Great Britain of the study of Arabic. [London, 1859]. 7 pp. Reprinted from the *Journal of the Royal Asiatic Society,* Vol. 16. (G.P.B., *i.e.* G. P. Badger). [*see* p.].

553 d
Sayce, A. H. An Assyrian grammar, for comparative purposes. London, Trübner, 1872. xvi, 188 pp.

554 l
Shakespear, J. A dictionary, Hindūstānī and English, and English and Hindūstānī, the latter being entirely new; fourth edition. London, Pelham Richardson for the author, 1849. xii pp., columns 1–2240, pp. 2241–414. [*see* q.].

555 a d
Smythe, P. E. F. W., Viscount Strangford. On the language of the Afghans. Part 1. [London, 1863]. 15 pp. Reprinted from the *Journal of the Royal Asiatic Society,* Vol. 20. [*see* 225].

556 a d
Snouck Hurgronje, C. Mekkanische Sprichwörter und Redensarten, gesammelt und erläutert. Haag, Martinus Nijhoff, 1886. [i], 144 pp. [*see* p.].

557 a d j
Socin, A. Arabic grammar, paradigms, litterature [*sic*], chrestomathy and glossary. Carlsruhe, Leipzig, H. Reuther; London, Williams & Norgate, 1885. xvi, 102, 192 pp., tables. (Porta linguarum orientalium, Vol. 4). (A. S.[ocin], Tüb.[ingen], 29 June 1885; in Arabic).

558 d
Spitta, W. Grammatik des arabischen Vulgärdialectes von Aegyptien. Leipzig, J. C. Hinrichs, 1880. xxxi, 519 pp., tables.

559 d
Steingass, F. The student's Arabic-English dictionary: companion volume to the author's English-Arabic dictionary. London, W. H. Allen, 1884. xvi, 1242 pp. wanting pp. 609–16.

560 a d
Stenzler, A. F. Elementarbuch der Sanskrit-Sprache: Grammatik, Text, Wörterbuch; dritte . . . Auflage. Breslau, Max Mälzer, 1875. [iv], 127 pp. [*see* p.].

561 a d
Summers, J. The rudiments of the Chinese language, with dialogues, exercises, and a vocabulary. London, Bernard Quaritch, 1864. [iv], 159 pp., front.

562 l
Ta'līm. Kitab ta'līm al-qirā'ah [A first reading book for children for use in the Protestant schools in Syria; Arabic text]. Beirut, 1866. 215 pp.

563 a e
Trumpp, E. Grammar of the Sindhi language, compared with the Sanskrit-Prakrit and the cognate Indian vernaculars. London, Trübner, [etc.], 1872. 16, 1, 540 pp., tables.

564 a d h
Wolff, J. F. A manual of Hebrew grammar, with points: or, a concise introduction to the Holy tongue. London, Liverpool, Dublin, James Cornish, [1839]. 48 pp.

565 a d (lower end-paper)
Yates, W. A dictionary in Sanscrit and English, designed for the use of private students and of Indian colleges and schools. Calcutta, Baptist Mission Press, printer, 1846. iv, 928 pp.

566 a d

— A grammar of the Sanscrit language, on a plan similar to that most commonly adopted in the learned languages of the West; second edition. Calcutta, Baptist Mission Press, printer, [etc.], 1845. xxi, 494 pp.

567 a d h

Yule, Sir H. and A. C. Burnell. Hobson-Jobson: being a glossary of Anglo-Indian colloquial words and phrases, and of kindred terms, etymological, historical, geographical, and discursive. London, John Murray, 1886. xlviii, 870 pp.

Linguistics – *Europe*

568 d

Anonymous. [Review of] Der germanische Urpsrung der lateinischen Sprache und des römischen Volkes, nachgewiesen von Ernst Jäkel. From the *Foreign quarterly review*, Vol. 10, no. 20, pp. 365–411. [*see* 218].

569 a d

— Vocaboli di prima necessitá e dialoghi famigliari ad uso degli studenti delle due lingue italiana e illirica; quarta edicione. Zara, Fratelli Battara Tip. Editori, 1870. 208 pp.

570

Alton, J. Beiträge zur Ethnologie von Ostladinien. Innsbruck, Verlag der Wagner'schen Universitäts-Buchhandlung, 1880. 68 pp. [*see* 571].

571

— Die ladinischen Idiome in Ladinien, Gröden, Fassa, Buchenstein, Ampezzo. Innsbruck, Verlag der Wagner'schen Universitäts-Buchhandlung, 1879. [iii], 376 pp.

572 a e

Ambra, R. d'. Vocabolario napolitano-toscano domestico di arti e mestieri. [Naples], A spese dell'autore, 1873. xi, 441 pp., fromt. (port.).

573 a i l

Andrews, J. B. Phonétique mentonaise. [Paris, 1883]. 6 pp. Reprinted from *Romania*, Vol. 12. [*see* p.].

574 a d h

Barrère, A. and C. G. Leland. eds. A dictionary of slang, jargon & cant embracing English, American, and Anglo-Indian slang, Pidgin English, Tinkers' jargon, and other irregular phraseology. [London], Ballantyne Press, printer for subscribers, 1889. 2 vols. No. 38 of an edition limited to 675 copies.

575 b d

Bartlett, J. R. A glossary of words and phrases usually regarded as peculiar to the United States; second edition, greatly improved and enlarged. Boston, Little, Brown; London, Trübner, 1859. xxxii, 524 pp. At head of title: Dictionary of Americanisms.

576 a

Blanchus, F. Dictionarium latino epiroticum una cum nonnullis usitatioribus loquendi formulis. Romae, Typis Sac. Congr. de Propag. Fide, 1635. [xvi], 224 pp.

577

Bomhoff, D. A new dictionary of the English and Dutch language; to which is added a catalogue of the most usual proper names and a list of the irregular verbs; carefully revised and considerably augmented; fourth edition. Nimmegen, J. F. Thieme, 1851. 2 vols.

578 k l

Bottarelli, F. Nouveau dictionnaire de poche: françois, italien et anglois; seconde édition. Nice, Société typographique, 1790–1. 3 vols.

579 k

Boyer, A. The Royal dictionary abridged in two parts. I, French and English – II, English and French; fifth edition, carefully corrected . . . London, printed for J. & J. Knapton, R. Wilkin, [etc.], 1728. 2 pts (in 1).

580 a d

Buehler, J. A. Grammatica elementara dil lungatg rhäto-romonsch per diever dils scolars en classas superiuras dellas scolas ruralas romonschas. I, part. Cuera, Squitchada en la Stamparia e de haver en la Libreria de L. Hitz, 1864. viii, 104 pp.

581 a

Carisch, O. Taschen-Wörtebuch der rhaetoromanischen Sprache in Graubünden . . . Chur, St Moritz, Hitz & Hail, 1887. xl, 214 pp. Reprint of the 1848 edition. wanting pp. 17–214. [*see* p.].

582 d i
Charnock, R. S. A glossary of the Essex dialect. London, Trübner, 1880. x, 64 pp.

583 h i
— Ludus patronymicus: or, the etymology of curious surnames. London, Trübner, 1868. xvi, 166 pp.

584 d i
— Patronymica cornu-britannica: or, the etymology of Cornish surnames. London, Longmans, Green, [etc.], 1870. xvi, 160 pp.

585 d i l
— Praenomina: or, the etymology of the principal Christian names of Great Britain and Ireland. London, Trübner, 1882. xvi, 128 pp.

586 a d h
Cleasby, R. An Icelandic-English dictionary based on the MS. collections of the late Richard Cleasby; enlarged and completed by Gudbrand Vigfusson; with an introduction and life of Richard Cleasby by George Webbe Dasent. Oxford, Clarendon Press, 1874. [4], cviii, 780 pp. With a second title-page dated 1869.

587 k
Clifton, C. E. and others. Manuel de la conversation et du style épistolaire . . . en six langues, français-anglais-allemand-italien-espagnol-portugais. Paris, Garnier Frères, [1859]. vi, [728] pp., tables.

588 l
Connellan, O. A practical grammar of the Irish language. Dublin, B. Geraghty, 1844. [iv], 120 pp.

589
Constancio, F. S. Novo diccionario portatil das lingoas portugueza e franceza; terceira edição; segunda parte: portuguez-francez. Paris, Rey e Gravier, 1830. [3], xxxvi, 560 pp.

590 d
Contopoulos, N. A lexicon of modern Greek-English and English-modern Greek. En Smyrnē, Typois B. Tatikidou; Typois B. Tatikianou, 1867-9. 2 vols. (in 1).

591
Corbella, C. [Dizionario italiano-inglese ed inglese-italiano: dizionario italiano-inglese. Milano, Tip. Fratelli Borroni, 1869]. 747-1614 pp. wanting title-page and pp. 769-84.

592 k
Cusani, F. and C. Grolli. Dizionario italiano-inglese ed inglese-italiano; compilato su quelli di Baretti e Meadows . . . seconda edizione. Genova, Presso Antonio Beuf, 1850. xl, 662, 66 pp.

593 a d h
Demattio, F. Grammatica della lingua provenzale con un discorso preliminare sulla storia della lingua e della poesia dei trovatori un saggio di componimenti lirici provenzali . . . Innsbruck, Libreria Accademica Wagner, 1880. [iv], 152 pp.

594
Diez, F. Etymologisches Wörterbuch der romanischen Sprachen; vierte Ausgabe, mit einem Anhang von August Scheler. Bonn, bei Adolph Maraus, 1878. xxvi, 820 pp.

595 a d
D'Orsey, A. J. D. Colloquial Portuguese: or, the words and phrases of every-day life . . . for the use of English tourists . . . second edition, considerably enlarged and improved. London, Trübner, 1860. viii, 126 pp.

596 a d
— A practical grammar of Portuguese and English, exhibiting, in a series of exercises in double translation, the idiomatic structure of both languages as now written and spoken, adapted to Ollendorf's system. London, Trübner, 1860. viii, 298 pp.

597
Forcellini, E. Totius latinitatis lexicon consiolio et cura Jacobi Facciolati, opera et studio Aegidii Forcellini, editit . . . auctarium denique et Horatii Tursellini de particulis, latinae orationis libellum etiam Gerrardi siglarium romanum et Gesneri indicem etymologicum, adjecit Jacobus Bailey. Londini, Sumptibus Baldwin et Cradock, [etc.], 1828. 2 vols., front., illus. [*see* q.].

598 l
Froelich, R. A. ed. Kurz gefasste tabellarisch bearbeitete Anleitung zur schnellen Erlernung der vier slawischen Hauptsprachen; ein Leitfaden, um in kurzer Zeit sich die böhmische, polnische, ilirische und russische Sprache vergleichungsweise eigen zu machen . . . Wien, Jos. Wenedikt's Witwe und Sohn, 1847. [iii], 152 pp.

599 a d

— Rěčnik ilirskoga i němačkoga jezika; sastavio ga Rud. V. Veselić. U Beču, A. A. Venedikt; Wien, Alb. A. Wenedikt, 1853–4. 2 vols. Title of vol. 2: Handwörterbuch der deutschen und ilirischen Sprache. Contents: 1, Ilirsko-neṁački dio – 2, Deutsch-ilirischer Theil.

600 d

— Theoretisch-praktische Grammatik der ilirischen Sprache; vierte Auflage, bearbeitet und mit Uebersetzungsstücken versehen von J. Macun. Wien, Albert Wenedikt, 1865. [viii], 322 pp., tables.

601 a d h

Gennadios, G. ed. Grammatikē tēs ellēnikēs glōssēs, suntachtheîsa men upo G. Gennadiou, kata diatagēn tēs kyvernēseōs, pros chrēstin tōn ellenikōn scholêion tou kratous, diaskeuastheisa de nun . . . upo G. Papasliōtou [Grammar of the Greek language . . . for use in Greek state schools . . . now revised by G. Papasliotes]. En Athēnais, 1857. x, 304 pp. [*see* 233].

602 a d

Gérard, F. and A. de Valdemare de Somow. L'interprète militaire en Orient français, russe, valaque et turc. Paris, Librairie d'Amyot, 1855. 156 pp.

603

[Gerardin, F. C. de la Place]. Néologie: or the French of our times; being a collection of more than eleven hundred words, either entirely new or remodernized . . . by Mme. Ve. D. G. London, P. Rolandi, 1854. [vi], 114 pp. [*see* 232].

604 k l

Gouldman, F. A copious dictionary in three parts . . . fourth edition. Cambridge, John Hayes, printer, 1678. 3 vols. (in 1).

605 l

Graglia, G. A. A new grammar of the Italian language, or a simple and easy plan; second edition. London, printed for Lackington, Allen, 1822. viii, 256 pp.

606 a d (vol. 2)

Hilpert, J. L. Dictionary of the English and German languages, founded upon the larger work. London, Franz Thimm, 1853. 2 vols.

607 a

Hossfeld, —. A new pocket dictionary of the English & Italian languages. London, Hirschfeld, n.d. 2 pts (in 1).

608 d

Janežič, A. Slovenisches Sprach- und Uebungsbuch für Anfänger; achte Auflage. Laibach, Zeschko & Till, 1872. [iv], 283 pp.

609

Kinloch, A. A compendium of Portuguese grammar; revised by A. J. Dos Reis. London, Edinburgh, Williams & Norgate, 1876. xii, 91 pp.

610 a d

Lastik, Iō. Stoicheiōdēs gallikē grammatikē, pros chrēsin tōn ellēnikōn scholeiōn. [Elementary French grammar, for use in Greek schools]. Athēnēsi, 1867. iv, 56 pp. [*see* 233].

611 k

Latham, R. G. A hand-book of the English language, for the use of students; fourth edition. London, Walton & Maberly, [etc.]; Longman, Green, [etc.], 1860. xxiv, 442 pp.

612

Lécluse, F. Manuel de la langue basque. Toulouse, Jn. Meu. Douladoure, [etc.]; Bayonne, L. M. Cluzeau, 1826. [viii], 224 pp. Preceded by Dissertation sur la langue basque, Toulouse, Vieusseux Père et Fils, 1826. 32 pp.

613 e (p. 5)

Loth, J. T. The tourist's conversational guide, in English, French, German and Italian. Edinburgh, Seton & Mackenzie, [etc.], 1867. vi, 83 pp. [*see* p.].

614

[Luciani, T.]. Sui dialetti dell'Istria. [Venezia], 1876. 29 pp. Reprinted from *Archivio veneto,* Vol. 11, part 2. [*see* p.].

615

Majer, M. Pravila kako izobraževati ilirsko narečje i u obče slavenski jezik . . . U Ljubljani, Jožef Blaznik, 1848. 130 pp. [*see* 608].

616 d

Mar, E. del. A complete & practical grammar of the Spanish language, as it is now spoken and written . . . with a comprehensive treatise on Castilian pronunciation;

sixth edition, carefully revised, enlarged, and improved. London, David Nutt, 1853. xii, 299 pp.

617

Meadows, F. C. New Spanish and English dictionary, in two parts. I, Spanish and English – II, English and Spanish; seventh edition. London, William Tegg, 1860. 2 vols. (in 1).

618 a

Michelsen, E. H. ed. The merchant's polyglot manual in nine languages. London, Longman, Green, [etc.], 1860. viii, 336 pp.

619 k l

Mottura, C. and G. Parato. Grammatica normale, teorica ed applicata, proposta alle scuole magistrali, tecniche e ginnasiali del regno; decimaterza edizione. Roma, Torino, [etc.], G. B. Paravia, 1875. 240 pp.

620

[Murray, J., the younger]. A handbook of travel-talk: being a collection of questions, phrases, and vocabularies in English, German, French, and Italian, intended to serve as interpreter to English travellers abroad, or foreigners visiting England; new edition, carefully revised. London, John Murray, 1871. xxi, 370 pp.

621 a d

Nacamulli, G. Elementi di grammatica greca ad uso degl' italiani. Corfu, Tipografia di G. Nacamulli, 1868. [ii], 231 pp., tables.

622 a d

O'Brien, J. Irish-English dictionary; second edition. Dublin, Hodges & Smith, 1832. lvi, 47 pp. [*see* 623].

623 a d

O'Brien, P. A practical grammar of the Irish language. Dublin, H. Fitzpatrick, 1809. x, 214 pp.

624 l

Ollendorff, H. G. A new method of learning to read, write, and speak a language in six months; adapted to the Spanish. London, Whittaker, Dulau, [etc.], 1858. vii, 579 pp., tables.

625 a d

Parčić, C. A. Grammatica della lingua slava (illirica). Zara, Spiridione Artale, 1873. viii, 200 pp., tables. [*see* 628].

626 k l

Pereira Coruja, A. A. Compendio da grammatica da lingua nacional. Rio de Janeiro, [Typographia de João Ignacio de Silva], 1866. 99 pp.

627

Peridēs, M. P. Encheiridion italikēs grammatikēs, pros chrēsin tōn tēn italikēn glōssan spoudazontōn neōn [Handbook of Italian grammar, for the use of youths studying the Italian language]. En Ermoupolei, 1868. 120 pp. [*see* 233].

628 a d j

Potočnik, B. ed. Grammatik der slowenischen Sprache. Laibach, gedruckt bei Joseph Blasnik, 1849. viii, 184 pp. (— Goernig, Trieste, 18 Oct. 1874).

629

Puoti, B. Vocabolario domestico napoletano e toscano; seconda edizione. In Napoli, dalla Stamperia del Vaglio, 1850. xvi, 696 pp.

630 a d l

Pyl, R. van der. A practical grammar of the Dutch language, containing an explanation of the different parts of speech, all the rules of syntax, and a great number of practical exercises. Rotterdam, printed for Arbon & Krap, 1819. [iv], 396 pp.

631 a d

Rask, E. A grammar of the Icelandic, or Old Norse tongue; translated from the Swedish . . . by George Webbe Dasent. London, William Pickering; Frankfort O/M, Jaeger's Library, 1843. viii, 272 pp., plate.

632 a d

— A short practical and easy method of learning the Old Norsk tongue, or Icelandic language . . . with an Icelandic reader, an account of the Norsk poetry and the sagas, and a modern Icelandic vocabulary for travellers, by H. Lund; second corrected edition. London, Franz Thimm, 1869. vi, 121 pp.

633 d

Razinger, A. and A. Zumer. Abecednik za slovenske ljudske šole . . . II, Neizpremenjeni natis. V Ljubljani, Kleinmayr & Fed. Bamberg, 1882. [i], 80 pp.

634 b

Richardson, C. A new dictionary of the English language, combining explanation with etymology, and illustrated by quotations from the best authorities; new edition. London, Bell & Daldy, [etc.], 1858. 2 vols. [*see* q.].

635

Rigutini, G. Vocabolario della lingua classica latina; aggiuntevi le corrispondenze italiane-latine, compilato per uso selle scuole; 2a edizione stereotipa. Firenze, G. Barbèra, 1881. 2 pts (in 1).

636 a d h l

Roget, P. M. Thesaurus of English words and phrases; classified and arranged so as to facilitate the expression of ideas and assist in literary composition; fourth edition. London, Longman, Brown, Green, [etc.], 1856. xl, 507 pp.

637

Rota, A. The Italian language, by Hossfeld's new method for self-tuition. London, Society for Promoting the Knowledge of Foreign Languages, [1887]. 13 pts. [*see* p.].

638 a d l

Saint-Loup, —. Grammaire des paresseux, grammaire française, complète sur un plan nouveau; septième édition. Paris, au Guide Rose des Etrangers, [187-]. xi, 114 pp.

639 a d

Sakellariou, A. A. Stoicheiōdēs ellēnikē grammatikē . . . pros chrēsin tōn en tois dēmotikois scholeîoîs didaskomenōn paidōn. [Elementary Greek grammar . . . for the use of children taught in elementary schools]. Athēnēsi, 1872. 78 pp. [*see* 233].

640 h l

Sauer, C. M. and G. Ferrari. Nuova grammatica tedesca, con temi, letture e dialoghi; edizione seconda. Eidelberga, Giulio Groos, [etc.]; Firenze, Ermmano Loescher, [etc.], 1874. xv, 422 pp.

641

Schlickum, P. P. A. Vocabolario italiano sistematico; italienisches Wörterbuch; zweite vermehrte Auflage. Poderhorn, 1875. v, 448 pp. wanting title page.

642 d

Schmitz, L. Grammar of the Latin language. London, Edinburgh, William & Robert Chambers, 1858. 318 pp.

643 i

Singer, I. Simplified grammar of the Hungarian language. London, Trübner, 1882. vi, 88 pp. (the publishers).

644 l

Smith, L. Guide to English and Portuguese conversation. Paris, Charles Hingray, [1863?]. [x], 356 pp., tables.

645 d

Solano Constancio, F. Novo diccionario critico e etymologico da lingua portugueza . . . precedido de huma introduccão grammatical; undecima edição. Paris, E. Belhatte, 1877. lii, 976 pp.

646 k

Spiers, A. General French and English dictionary, newly composed from the French dictionaries of the French Academy, Laveaux, Boiste, Bescherelle, etc., from the English dictionaries of Johnson, Webster, Richardson, etc.; eleventh edition. London, Whittaker, 1859. 2 vols.

647 a d h

Szoelloesy, J. N. von. Sprachlehre, um nach Ollendorff's Methode mittelst selbstunterricht in der kürzest möglichen Zeit: französisch, deutsch, englisch, italienisch, russisch . . . geläufig Sprechen und verstehen zu Lernen. Klausenburg, bei Barra's Witwe und Stein, 1850. 14 pts (in 1).

648 a d l

Tergolina, V. di. The door to the Italian grammar. London, Longman, Green, [etc.]. 1869. 52 pp., tables.

649 a d h

Valpy, R. The elements of Greek grammar; a new edition. London, Longman, Brown, [etc.], 1858. [ii], 214 pp., front. [*see* 233].

650 a d l

Vieyra, A. A dictionary of the Portuguese and English languages, in two parts: Portuguese and English, and English and Portuguese . . . a new edition . . . by A. J. da Cunha. Part I, Portuguese and English. London, Longman, Orme, [etc.], 1840. [viii], 702 pp. Dated by Burton: Bombay 1843.

651 l
— A grammar of the Portuguese language; to which is added a copious vocabulary, and dialogues, with extracts from the best Portuguese authors; twelfth edition. London, Dulau, 1858. viii, 443 pp.
652 l
— Novo diccionario portatil das linguas portugueza e ingleza em duas partes: portugueza e ingleza – ingleza e portugueza; resumido do diccionario de Vieyra; nova edição revista e consideravelmente augmentada por J. P. Aillaud. Paris, em Casa de Va J.-P. Aillaud, Monlon, 1862. 2 vols. (in 1).
653 a d
Vlachos, A. Elementar-Grammatik der neugriechischen Sprache; zweite verbesserte Auflage. Leipzig, F. A. Brockhaus, 1871. ix, 86 pp. [*see* 621].
654 a d
— Neugriechische Chrestomathie oder Sammlung von Musterstücken der neugriechischen Schriftsteller und Dichter; zusammengestellt und mit erklärenden Anmerkungen versehen. Leipzig, F. A. Brockhaus, 1870. vii, 186 pp. [*see* 621].
655 k
Walker, J. A critical pronouncing dictionary, and expositor of the English language: . . . to which are prefixed, principles of English pronunciation . . . sixteenth edition. London, T. Cadell & W. Davies, [etc.], 1816. [2], 91, 602 pp.
656 d
— — a new edition. Glasgow, James Lumsden, 1843. [i], 84, 588, [ii], 66 pp., front. (port.).
657 d
Worcester, J. E. Dictionary of the English language . . . London, Sampson Low, [etc., 1859]. lxviii, 1786 pp., illus. [*see* q.].
658 a d
Wright, T. Dictionary of obsolete and provincial English, containing words from the English writers previous to the nineteenth century, which are no longer in use, or are not used in the same sense; and words which are now used only in the provincial dialects. London, Bell & Daldy, 1869. 2 vols.
659
Xylander, J. L. von. Die Sprache der Albanesen oder Schkipetaren. Frankfurt am Main, in der Andreaischen Buchhandlung, 1835. xvi, 320 pp., tables.
660
Zotti, R. A general table of the Italian verbs, regular and irregular, by which, the formation of any tense or person required, may be immediately found; after the table of the French verbs by R. Juigné. London, executed and sold by R. Zotti, [*c.* 1873?]. 1 sheet. [*see* p.].

661-1105 Literature

661
Anonymous. Album da raspasiada: eróticos feitos e, na maior parte, improvisados por B. M. F. Bruxellas, Typographia de Richard Brenneke, 1864. 120 pp. [*see* 249].

662 i
— John Dudley: a tragedy for the stage and closet, by Scriptor Ignotus. London, Reeves & Turner, 1886. vii, 111 pp. (24 June 1886).

663
— Lo spettro dell'antenata: leggenda allemanna del secolo XII; traduzione di C.M.P. Londra, P. Rolandi, 1865. 64 pp.

664
— The O'Donnells of Innismore: or, the two Marys, by the authoress of "Grace O'Halloran". London, Richardson, 1864. [xii], 218 pp.

665
— Schattenspiel: aufgeführt in Abbazia am 7. März 1888, anlässlich der Gründung der Section "Quarnero" des "Union-Yacht-Club". [Fiume, Tipo-Litographische Anstalt des Emidio Mohovich], 1888. 79 pp., illus. [*see* q.].

666 a
— Skýrslur og reikníngar hins Íslenzka bókmentafèlags, 1870-1871. Kaupmannahöfn, S. L. Möllers, 1871. xlvi, 43 pp. [*see* p.].

667 a d h
Abreu Medeiros, F. L. d'. Curiosidades brasileiras. Rio de Janeiro, Eduardo & Henrique Laemmert, 1864. 2 vols. plates.

668 a
[Adams, W. B.]. An appeal to the British nation against the government's Eastern policy, by Junius Redivivus. Zara, Tipografia G. Woditzka, 1884. 79 pp. [*see* p.].

669 a
Aeschylus. The new readings contained in Hermann's posthumous edition of Aeschylus; translated and considered by George Burges. London, Henry G. Bohn, 1853. [iii], 156 pp. (Bohn's classical library).

670 a d
— The tragedies of Aeschylus; literally translated with critical and illustrative notes . . . by Theodore Alois Buckley. London George Bell, 1876. xx, 234 pp., front. (port.). (Bohn's classical library).

671 k l
Alencar, J. M. de. Cinco minutos – a viuvinha. Rio de Janeiro, B. L. Garnier, 1865. [iii], 216 pp. (IB's copy).

672 d k l
— Iracema, lendo do Ceará. Rio de Janeiro, Vianna & Filhos, 1865. vi, 202 pp. (IB's copy; all annotations etc. are by her except pp. 173, 181, 183, 190).

673 d
— O Guarany: romance brasileiro; segunda edicçâo. Rio de Janeiro, B. L. Garnier, [1857?]. 2 vols.

674 a d
Ammianus Marcellinus. The Roman history of Ammianus Marcellinus . . . translated by C. D. Yonge. London, Henry G. Bohn, 1862. vii, 646 pp. (Bohn's classical library).

675
Andersen, H. C. H. C. Andersen's Sämmtliche Märchen; dreiundzwanzigste Auflage. Leipzig, Ed. Wartig, 1882. vii, 727 pp., front., illus., plates.

676 a d
Apuleius, L. The works of Apuleius . . . a new translation . . . London, George Bell, 1878. ix, 533 pp., front. (Bohn's classical library).
677 d
Arabian Nights – Arabic. Tausend und eine Nacht, Arabisch; nach einer Handschrift aus Tunis; herausgegeben von Dr. Maximilian Habicht; [Arabic text]. Breslau, mit Koniglichen Schriften; Josef Max, Ferdinand Hirt, 1825–43. 12 vols.
678 d
— — Alf lailah wa alilah [The thousand and one nights; Arabic text]. Būlāq, 1251 A.H. [1836 A.D.]. 2 vols.
679
— — The alif laila, or Book of the thousand nights and one night . . . now for the first time, published complete in the original Arabic, from an Egyptian manuscript . . . edited by W. H. Macnaughten. Calcutta, 1839–42. 4 vols.
680 a d j
— English. Arabian tales: or, a continuation of the Arabian nights entertainments . . . translated from the French into English, by Robert Heron. Edinburgh, printed for Bell & Bradfute, [etc.], 1792. 4 vols., plates. (Leonard Smithers, Sheffield, 15 Oct. 1885). *wanting* Vol. 2.
681 a d
— — The Arabian nights entertainments, carefully revised, and occasionally corrected from the Arabic; to which is added, a selection of new tales; now first translated from the Arabic originals; also an introduction and notes . . . by Jonathan Scott. London, printed for Longman, Hurst, [etc.], 1811. 6 vols. fronts.
682 a d g h j
— — The thousand and one nights, commonly called, in England, the Arabian nights' entertainments; a new translation from the Arabic, with copious notes, by Edward William Lane. London, Charles Knight, 1839–41. 3 vols., illus. (A. J. Cotheal, New York, 9 Feb. 1885; E. F. Marriott, Bombay, 18 Feb. 1885).
683 d
— — The Arabian nights' entertainments; translated by the Reverend Edward Forster; new edition . . . explanatory and historical introduction, by G. Moir Bussey. London, H. Washbourne, 1853. xxxviii, 490 pp. front., plates.
684
— — The Arabian nights' entertainments; translated from the Arabic. Lucknow, re-printed at the "Newul Kishore" Press, 1879–80. 4 parts (in 2 vols).
d
Second copy of Parts 3–4 bound in one volume, and partially interleaved; part 4, pp. 1–5 'The adventures of the Caliph Haroun Alraschid' heavily annotated.
685 a d h
— — The book of the thousand nights and one night; now completely done into English prose and verse, from the original Arabic, by John Payne. London, printed for the Villon Society by private subscription and for private circulation only, 1882–9. 13 vols. Vols [10–12] entitled: Tales from the Arabic of the Breslau and Calcutta (1814–18) editions of the Book of the thousand nights and one night. Vol. [13] entitled: Alaeddin and the enchanted lamp.
Upper cover of each volume lettered in gold: R.F.B.
686 d l
— French. Les mille et une nuits: contes arabes; traduits en Français par Galland; nouvelle édition . . . par M. Édouard Gauttier. Paris, J. A. S. Collin de Plancy, 1822–3. 7 vols.
687 d
— — Les mille et une nuits: contes arabes traduits par Galland; édition illustrée; revue et corrigée . . . augmentée d'une dissertation . . . par M. le Baron Silvestre de Sacy. Paris, Ernest Bourdin, [1840]. 3 vols., fronts, illus., plates.
688 a d
— — Contes inédits des mille et une nuits, extraits de l'original arabe par M. J. de Hammer . . . traduits en Français par M. G.-S. Trébutien. Paris, Dondey-Dupré, 1828. 3 vols.
689
— German. Die Gedichte des 'Alkama Alfahl, mit Anmerkungen; herausgegeben von Albert Socin. Leipzig, F. C. W. Vogel, 1867. viii, 42, [24] pp.
690 a d h
— — Tausend und eine Nacht: Arabische Erzählungen; zum erstenmale aus dem

Urtexte vollständig und treu übersetzt von Dr. Gustav Weil. Stuttgart, Rieger, 1872. 4 vols. (in 2), fronts., illus.

691 l

— Icelandic. Iúsund og ein nótt: Arabiskar sögur. Kaupmannahöfn, Páll Sveinsson, 1857–64. 4 vols.

692 h

— Italian. Le mille e una notte: novelle arabe. Roma, Edoardo Perino, 1882. 542 pp., illus.

693

— Urdu. Alif laila Urdu; [translated by Abdul Karim]. Kanpor, Mutba Mustafai, 1860. 4 vols. (in 1).

694 d (Vol. 1)

— — Hazar dastan . . . Alif laila nasir. Kawnpore, Mutba Munshi Nawal Kishor, 1300 A.H. [1883 A.D.]. 4 vols. (in 1), illus., plates.

695

— — Turjuma Alif laila, ba-zaban Urdu; [translated by Abdul Karim]. Kanpor, Mutba Mustafai, 1300 A.H. [1883 A.D.]. 4 vols. (in 1). illus.

696

— Selections. Arabic. The Arabian nights entertainments, in the original Arabic; published under the patronage of the College of Fort William, by Shuekh Uhmud bin Moohummud Shirwanee Ool Yumunee. Calcutta, P. Pereira, 1814–18. 2 vols. (in 1).

697

— — — Contes arabes, extraits des manuscrits de la Bibliothèque Nationale par Florence Groff. Paris, Ernest Leroux, 1888. [86] pp.

698 a d

— — — Histoire d'Alâ al-Dîn: ou la Lampe marveilleuse; texte arabe publié avec une notice sur quelques manuscrits des Mille et une nuits, par H. Zotenberg. Paris, Imprimerie Nationale, 1888. [iv], 70 pp., and Arabic text, 90 pp., plate.

699

— — Dutch. Oostersche Vertellingen uit de Duizend-en-één-nacht; naar de Hoogduitsche Bewerking van M. Claudius, voor de Nederlandsche jeugd uitgegeven door J. J. A. Goeverneur. Groningen, J. B. Wolters, [*c.* 1860]. [iii], 282 pp., front.

700

— — English. Caliphs and sultans, being tales omitted in the usual editions of the Arabian nights entertainments; re-written . . . by Sylvanus Hanley. London, L. Reeve, 1868. viii, 263 pp.

701 d i

— — — The new Arabian nights, selected tales, not included by Galland or Lane; [translated and edited by K. F. Kirby]. London, W. Swan Sonnenschein, [1882]. xii, 390 pp.

702 a d g h

— — — The book of Sindi-bad . . . from the Persian and Arabic; with introduction . . . by W. A. Clouston. [Glasgow], privately printed, 1884. lvi, 385 pp., table.

703 d

— — French and Turkish. Histoire de Calife le pêcheur et du calife Haroun er-Rechid . . . text turc . . . et de la traduction française, par Charles Clermont-Ganneau. Jérusalem, Typographie de Terre Sainte, 1869. 128 pp.

704 i

Araujo, J. de. Occidentaes. Porto, Lugan & Genelioux, 1888. 150 pp.

705

— Poetas mortos, consagracões. Porto, Typographia occidental, 1888. 29 pp.

706 a d h j

Aravantinos, P. Paroimiastēriou: ē sullogē paroimiōu en hrēsei ousou para tois Ēpeirōtais, met' anaptyxseōs tēs ennoias rautōn kai parallēlismou pros tas archaias. [Paroimiastēriou: or a collection of proverbs in use by the Epeirotes with an explanation of their meaning, and a comparison with ancient ones]. En Iōanninois, Typografeion Dōdōnēs, 1863. 183 pp. (William Sterling Maxwell, Dublane, 2 March and 29 Nov. 1873; C. Mayreder, Vienna, 13 May 1878).

707 d

Arbuthnot, F. F. Arabic authors: a manual of Arabian history and literature. London, William Heinemann, 1890. xiv, 247 pp.

708

[—]. Early ideas: a group of Hindoo stories; collected and collated by Anaryan. London, W. H. Allen, 1881. 158 pp.

709 a d h
— Persian portraits: a sketch of Persian history, literature, and politics. London, Bernard Quaritch, 1887. xii, 170.

710 d l
Ariosto, L. Orlando furioso; translated from the Italian; with notes, by John Hoole; second edition. London, George Nicol, 1785. 5 vols. plates.

711 a d j
— Orlando furioso in English heroical verse, by Sr John Harington. London, G. Miller, printer for J. Parker, 1634. [xvi], 432 pp., illus. (Wyman & Sons, London, 22 April 1885).

712 d
— Orlando furioso illustrato da Gustavo Doré; con prefazione di Giosuè Carducci. Milano, Fratelli Treves, 1881. xx, 642 pp., front., illus., plates. Upper cover lettered: Ariosto. Isabel Burton. [*see* q.].

713 a d
Aristophanes. The comedies of Aristophanes; a new and literal translation from the revised text of Dindorf; with notes . . . by William James Hickie. London, George Bell, 1877–8. 2 vols. front. (port.). (Bohn's classical library).

714 a d
Aristotle. History of animals, in ten books; translated by Richard Cresswell. London, Henry G. Bohn, 1862. ix, 326 pp. (Bohn's classical library).

715
— The metaphysics of Aristotle, literally translated from the Greek, with notes, analysis . . . by the Rev. John H. M'Mahon. London, George Bell, 1876. xcvi, 445 pp. (Bohn's classical library).

716
— The Nicomachean ethics . . . translated, with notes . . . by R. W. Browne. London, George Bell, 1877. lxxxi, 348 pp. (Bohn's classical library).

717 e
— The Organon, or logical treatises, with the introduction of Porphyry; literally translated; with notes . . . by Octavius Freire Owen. London, George Bell, 1877–8. 2 vols. (Bohn's classical library).

718
— The politics and economics of Aristotle, translated, with notes, original and selected, and analyses, to which are prefixed, an introductory essay and life of Aristotle, by Dr Gillies, by Edward Walford. London, George Bell, 1876. lxxx, 338 pp. (Bohn's classical library).

719
— Treatise on rhetoric; literally translated with Hobbes' analysis . . . by Theodore Buckley. London, George Bell, 1878. iv, 500 pp., front. (port.). (Bohn's classical library).

720
Arnold, Sir E. Hero and Leander; from the Greek of Musaeus. London, Casell, Petter & Galpin, [1873]. [i], 32 pp., front.

721
— Indian poetry; third edition. London, Trübner, 1884. viii, 270 pp.

722 l
— The secret death (from the Sanskrit) with some collected poems by Edwin Arnold; third edition. London, Trübner, 1885. viii, 406 pp.

723 k l
— The song celestial or Bhagavad-gîtâ; translated from the Sanscrit text. London, Trübner, 1885. xiv, 173 pp.

724 a
Arnold, F. G. Septem Mo'allakât carmina antiquissima arabum; textum ad fidem optimorum codd. et editt. recensuit . . . annotationes criticas adiecit D. Fr. Aug. Arnold. Lipsiae, Guil. Vogelii, filii, 1850. x, 64, [220] pp. Also carries autograph signature of W. G. Palgrave.

725 a
Arpagaus, G. Fablas e novellas; dedicadas alla giuventegna romonscha. Cuera, Stampadas dals frars Casanova, 1878. iii, 105 pp.

726 a d
Athenaeus. The deipnosophists or banquet of the learned of Athenaeus; literally translated by C. D. Yonge. London, Henry G. Bohn, 1854. 3 vols. (Bohn's classical library).

727 **d**
Aurelius Antoninus, M. The thoughts of the Emperor M. Aurelius Antoninus; translated by George Long; second edition. London, George Bell, 1877. [vii], 210 pp. (Bohn's classical library).
728 **d**
Ausonius, D. M. Decimi Magni Ausonii burgidalensis opuscula; recensuit Rudolfus Peiper. Lipsiae, B. G. Teubner, 1886. cxxviii, 556 pp., plate.
729 **a d h**
— Oeuvres complètes d'Ausone; traduction nouvelle, par E.-F. Corpet. Paris, C. L. F. Panckoucke, 1842–3. 2 vols.
730
Avitabile, F. Alla nobile Dama Lady Elisa Otway queste poche rime di eletti autori Filomena Avitabile. Napoli, pe' Tipi di Vincenzo Marchese; 1875. 16 pp. [*see* p.].
731 **a d i**
Baddeley, W. St C. Bedoueen legends and other poems. London, Robson & Kerslake, 1883. xv, 144 pp. (22 June 1883).
732 **d i**
— Dramatic and narrative sketches. London, Robson & Kerslake, 1885. [v], 219 pp. (28 Nov. 1884).
k
Another copy.
733 **k l**
— Tennyson's grave. London, William Heinemann, 1893. 16 pp. [*see* p.].
734 **a d**
Baitāl-Pachīsī. The Baitál Pachísí: or, twenty-five tales of a demon; a new edition of the Hindi text, with . . . literal English interlinear translation . . . by W. Burckhardt Barker; edited by E. B. Eastwick. Hertford, Stephen Austin, 1855. xi, 369 pp.
735 **k**
Baldegg, H. von. ed. Marienkrone: Perlen und Blüthen aus dem deutschen Dichtergarten. Solothurn, Scherer, 1858. xiv, 160 pp.
736
Balzac, H. de. Physiologie du mariage ou méditations de philosophie éclectique sur le bonheur et le malheur conjugal; nouvelle édition. Paris, Charpentier, 1847. [iii], 376 pp.
737 **k**
Barlow, G. The crucifixion of man: a narrative poem. London, Swan Sonnenschein, 1893. xix, 231 pp., front. (port.).
738 **a d j**
Basile, G. B. – English. The Pentamerone: or the story of stories; translated from the Neapolitan by John Edward Taylor; second edition. London, David Bogue, 1850. xvi, 404 pp., front., plates. (Eliza Otway, undated; Ulrico Hoepli, bookseller, Milan, 3 May 1886).
739 **a e h**
— German. Der Pentamerone or, das Märchen aller Märchen; aus dem neopolitanischen übertragen von Felix Liebrecht. Breslau, Josef Max, 1846. 2 vols.
740
— Italian. Il Pentamerone . . . ouero lo cunto de li cunte; trattenemiento di li Peccerille di Gian Alesio Abbattutis; nouamente restampato . . . al illustrissimo . . . Giuseppe Spada. Roma, nella Stamperia di Bartolomeo Lupardi, 1679. [viii], 636 pp.
741 **a**
— — Il conto de' conti, trattenimento a' fanciulli; trasportato dalla Napoletana all' Italiana favella . . . Napoli, Libreria di Cristoforo Migliaccio, [etc.], 1754. [iv], 264 pp., illus.
742 **a d**
— — Il Pentamerone: overo lo cunto de li cunte; trattenemiento de li Peccerille di Gian Alesio Abbattutis. Napoli, Guiseppe-Maria Porcelli, 1788. 2 vols. (Collezione . . . poemi in lingua napolitana, Vols. 20–1).
Another copy.
743 **e (p. 19)**
— — Il conto de' conti . . . transportato dalla Napolitana all'Italiana favella. Napoli, nella Tipografia Michele Migliaccio, 1821. [i], 264 pp., illus.
744 **e (p. 215)**
— — Il conto de' conti; trattenimento a' fanciulli; nuova edizione. Napoli, Gennerao Cimmaruta, 1863. 216 pp.

745 d
Benizelos, I. Paroimíai dēmōdeis sullegeīsai kaì ermēneutheīsai 'upò I. Benizélou . . . ekdosis deutéra epēuxēménē kaì diōrthōménē [Popular proverbs collected and interpreted by I. Benizelos; second edition enlarged and corrected]. En 'Ermoupólei, ek toū tupografeíou tēs "Patrídos", 1867. viii, 360 pp. [*see* 706].
746 k
Berquin, A. L'ami des enfans; nouvelle édition. Lyon, Perisse Frères, [etc.], 1835-8. Vols. 1-4, 7.
747 l
Besant, W. The art of fiction: a lecture delivered at the Royal Institution on Friday evening, April 25, 1884, with notes and additions. London, Chatto & Windus, 1884. 38 pp. imperfect. [*see* p.].
748 a d
Bídpáí. The fables of Pilpay; revised edition. London, Frederick Warne, [1886]. xviii, 274 pp., illus. (Chandos classics).
749 a d h
– Johannis de Capua Directorium vitae humanae . . . version latine du livre de Kalilah et Dimnah, publiée et annotée par Joseph Derenbourg, 1er fascicule. Paris, F. Vieweg, 1887. [iii], 240 pp.
750 a d h j
— Kalilah and Dimmnah: or the fables of Bidpai . . . with an English translation by I. G. N. Keith-Falconer. Cambridge, University Press, 1885. lxxxvii, 320 pp. (Henry Bradley, [London], 23 March 1885; A. Socin, Tübingen, 30 April 1885).
751
[Bingham, W. R.]. The field of Ferozeshah, in two cantos, with other poems, by a young soldier, who fought in that glorious campaign. London, Charles Edward Bingham, 1848. [xi], 84 pp. [*see* p.].
752 d i
[Bocayuva, Q. preface]. Lyrica nacional. Rio de Janeiro, Typographia do Diario do Rio de Janeiro, 1862. v, 125 pp. (Bibliotheca brasileira, 1). [*see* 252].
753 a d
Boccaccio, G. The Decameron: or ten days' entertainment; new edition . . . by G. Standfast. London, Charles Daly, [1845]. [v], 500 pp., front., plates.
754 a d h
— Il Decameron. Leipzig, F. A. Brockhaus, 1877. 2 vols. (in 1). (Biblioteca d'autori italiana).
755 k
[Boehl von Faber, C. F. J.]. Un verano en bornos, novella de costumbres, por Fernan Caballero. Madrid, Francisco de P. Mellado, 1858. x, 264. (Professor A. P. Corrêa Pomentel, 21 June 1857).
756 a
Borelli, G. Saggio di scritti fantastico-umoristici. Trieste, Tipografia Morterra, 1880. [iv], 124 pp. [*see* p.].
757 a d
Bouterwek, F. History of Spanish literature; translated from the original German by Thomasina Ross. London, David Bogue, 1847. xiv, 450 pp., front. (port.).
758 d
— History of Spanish and Portuguese literature; translated from the original German by Thomasina Ross. Vol. II, Portuguese literature. London, Boosey, 1823. 12, 406 pp.
759 k
Browne, M. A. The birth-day gift. London, Hamilton, Adams, 1834. iv, 184 pp.
760 g h
Buchanan, R. Balder the beautiful: a song of divine death. London, William Mullan, 1877. xvi, 312 pp.
761 d
Byron, G. G. N. 6th Baron Byron. The works of Lord Byron, including the suppressed poems; also a sketch of his life, by J. W. Lake. Philadelphia, R. W. Pomeroy & Henry Adams, 1829. xxxix, 716 pp., front., plates.
762 d
Caesar, C. J. C. Julii Caesaris Commentarii de bello gallico. London, Edinburgh, William & Robert Chambers, 1857. xvi, 240 pp., front. (map), illus. (Chambers's educational course, classical section). [*see* 642].
763 a d f l
— Commentaries on the Gallic and civil wars . . . literally translated. London, George

Bell, 1876. iv, 572 pp., front. (port.). (Bohn's classical library).

764 h l

Calderon de la Barca, P. Las comedias . . . corregidas y dadas á luz por Juan Jorge Keil. Leipsique, Ernesto Fleischer, 1827–30. 4 vols.

765 a d

Camoens, L. de. – complete works – Portuguese. Obras . . . terceira edição. Paris, P. Didot Senior; Lisboa, Viuva Bertrand e Filhos, 1815. 5 vols. (in 6), fronts., plates, map.

766 a d

— — — Obras completas . . . correctas e emendadas . . . de J. V. Barreto Feio e J. G. Monteiro. Hamburgo, Langhoff, 1834. 3 vols. wanting Vol. 1.

767 a d

— — — Obras . . . precedidas de um ensaio biographico . . . com algumas composições ineditas do poeta pelo Visconde de Juromenha. Lisboa, Imprensa Nacional, 1860–9. 6 vols. (in 5), front. (port.), plates (Vol. 6).
Inserted in Vol. 2: Sechs Sonette von Luis de Camões, Abschrift von Wilh. Storck.

768 a d l

— — — Obras completas . . . edição critica. Porto, Imprensa Portugueza, 1873–4. 2 vols.

769 i

— Dramatic poems – German. Dramatische Dichtungen; zum ersten Male deutsch von Wilhelm Storck. Paderborn, Ferdinand Schöningh, 1885. [viii], 426 pp. (Sämmtliche Gedichte, 6). (Wilhelm Storck, Munster i / W., 4 Nov. 1884).
Two other copies.

770 a d

— Elegies – German. Buch der Elegieen, Sestinen, Oden und Octaven; zum ersten Male deutsch von Wilhelm Storck. Paderborn, Schöningh, 1881. xvi, 434 pp. (Sämmtliche Gedichte, 3).

771 a d j

— Idylls – German. Sämmtliche Idyllen; zum ersten Male deutsch von C. Schlüter und W. Storck. Münster, Adolph Russell, 1869. xxiii, 254 pp.
Another copy.

772 i

— — — Buch der Canzonen und Idyllen; Deutsch von Wilhelm Storck; zweite und verbesserte Auflage. Paderborn, Ferdinand Schöningh, 1882. xiii, 442 pp. (Sämmtliche Gedichte, 4). (Wilh. Storck).

773 a d

— The Lusiads – English. The Lusiad, or, Portugals historical poem . . . now newly put into English by Richard Fanshaw Esq. London, printed for Humphrey Moseley, 1655. [xxii], 224 pp., front. (port.), plate.

774 j

— — — The Lusiad: or, the discovery of India . . . translated from the original Portuguese of Luis de Camoëns by William Julius Mickle; third edition. London, printed for T. Cadell Jun. & W. Davies, 1798. 2 vols. (Leonard Smithers, Sheffield, 17 May 1887).

775 a d

— — — The Lusiad: or, the discovery of India: an epic poem; translated from the Portuguese of Luis de Camoēns . . . by William Julius Mickle; a new edition. London, printed for Joseph Harding, 1807. 3 vols., fronts., plates.

776 a d

— — — The Lusiad, an epic poem, by Luis de Camoens translated from the Portugueze by Thomas Moore Musgrave. London, John Murray, 1826. xxiii, 585 pp.

777 a d

— — — The Lusiad of Luis de Camoens. Books I. to V., translated by Edward Quillinan with notes by John Adamson. London, Edward Moxon, 1853. xv, 207 pp., front. (port.).

778 a d

— — — The Lusiad of Luis de Camoens, closely translated . . . by . . . Sir T. Livingston Mitchell. London, T. & W. Boone, 1854. xxix, 310 pp., front. (port.), plate.

779 a d h i l

— — — The Lusiads of Camoens; translated into English verse by J. J. Aubertin. London, C. Kegan Paul, 1878. 2 vols. fronts. (ports.), plates, map. (J. J. Aubertin).

780 a g

— — — The Lusiads of Camoens; translated into English verse by J. J. Aubertin;

second edition. London, Kegan Paul, Trench, 1884. 2 vols. fronts. (holograph poem by Aubertin, dated Washington, 6 Dec. 1886).
Another copy.

781 a d
— — — The Lusiad of Camoens; translated into English Spenserian verse by Robert ffrench Duff. Lisbon, London, Chatto & Windus, [etc.], 1880. xlviii, 509 pp., front. (port.), plates.

782 a d
— — French. Les Lusiades: ou, les Portugais, poëme en dix chants par Camoens; traduction de J. B. J. Millié, revue, corrigée . . . par M. Dubeux. Paris, Charpentier, 1862. [3], lx, 368 pp.

783 a d
— — Portuguese. Os Lusiadas: poema epico de Luis de Camões; restituido a' sua primitiva linguagem . . . por José da Fonseca. Paris, Europea de Baudry, [etc.], 1846. xxxiv, 586 pp., front. (port.).

784
— — — Os Lusiadas; edição revista e prefaciada por Theophilo Braga. Lisboa, Pereira & Amorim, 1881. 2 vols. (in 1).

785 a d
— — Spanish. Lusiadas de Luis de Camoens . . . comentadas por Manuel de Faria i Sousa. Madrid, Iuan Sanchez, 1639. 2 vols. (*i.e.* 4 parts)., illus. Title on spine: Obras de Camoes I, [. . .] II. [*see* q.].

786 a d j
— Songs. German. Sämmtliche Canzonen des Luis de Camoens; zum ersten Male deutsch von Wilhelm Storck. Paderborn, Ferdinand Schöningh, 1874. xxiii, 156 pp. (Wilh. Storck, Münster i. W. 16 Dec. 1884).

787 a d
— Songs and letters. German. Buch der Lieder und Briefe; zum ersten Male deutsch von Wilhelm Storck. Paderborn, Ferdinand Schöningh, 1880. xxix, 408 pp.

788 a d h i l
— Sonnets. English. Seventy sonnets of Camoens; Portuguese text and translation, with original poems, by J. J. Aubertin. London, C. Kegan Paul, 1881. xxiii, 253 pp. (5 March 1881).
d k
Another copy.

789 a d
— — German. Sonette von Luis Camoens; aus dem Portugiesischen von Louis von Arentschildt. Leipzig, F. A. Brockhaus, 1852. xx, 288 pp. [*see* 786].

790 a d h
— — — Buch der Sonette; Deutsch von Wilhelm Storck. Paderborn, Ferdinand Schöningh, 1880. xxxi, 439 pp. (Sämmtliche Gedichte, 2).

791 a d
— Selections. Poems from the Portuguese of Luis de Camoens; with remarks on his life and writings, notes, &c. &c. by Lord Viscount Strangford; fourth edition. London J. Carpenter, 1805. [iii], 160 pp., front. (port.).

792 a d
— — Rimas varias de Luis de Camoens . . . commentadas por Manuel de Faria y Sousa. Lisboa, Imprenta de Theotonis Damasco de Mello; Imprenta Craesbeeckiana 1685-9. 2 vols. [*i.e.* 5 parts]. Title on spine: Obras de Camoes, III, [. . .] IV. [*see* q.]

793 d h i
— Bibliography. Bibliographia camoniana, por Theophilo Braga. Lisboa, Imprensa de Christovão A. Rodrigues, 1880. 255 pp. (Antonio Augto. de Carvo. Monteiro Lisbonne, 20 Feb. 1884).

794 d
— — Camoens in Deutschland: bibliographische Beiträge zur Gedächtnissfeier des Lusiadensängers [von] Wilhelm Storck. Koloszvár, Acta Comparationis Litterarum Universarum, Universitätsbuchdruckerei Johann Stein, 1879. 48 pp. [*see* 790].

795 k
— — Collecção camoneana de José do Canto; tentativa de um catalogo methodico e remissivo. Lisboa, Imprensa Nacional, 1895. xi, 361 pp., front.

796 a d g h
— Biography and criticism. Memoirs of the life and writings of Luis de Camoens, by John Adamson. London, printed for Longman, Hurst, [etc.], 1820. 2 vols. fronts. (ports.), illus., plates.

797 a d
— — Censura das Lusiadas, por José Agostinho de Macedo. Lisboa, Impressão Regia, 1820. 2 vols.

798 a d
— — Estudo moral e politico sobre Os Lusiadas, por José Silvestre Ribeiro. Lisboa, Imprensa Nacional, 1853. xi, 239 pp.

799 a d i
— — Camões e Os Lusiadas, por Joaquim Nabuco. Rio de Janeiro, Imperial Instituto artistico, 1872. 286, viii pp. (London, 5 July 1882).

800 d
— — Camoens et Les Lusiades: étude biographique, historique et littéraire . . . par Clovis Lamarre. Paris, Didier, 1878. vii, 614 pp.

801 d h
— — Luigi di Camoens 300 anni dopo la sua morte, [by] Raffaele Cardon. Roma, Tipografia Barbèra, 1881. 173 pp.

802 a i
— — Luis de Camoens Leben; nebst geschtlicher Einleitung von Wilhelm Storck. Paderborn, Ferdinand Schöningh, 1890. xvi, 702 pp.

803 a d
Carlyle, J. D. Specimens of Arabian poetry. Cambridge, John Burges, printer to the University, 1796. ix, 181 pp. *and* Arabic text, 73 pp.

804 i l
Castilho, J. de. D. Ignez de Castro: drama em cinco actos. Rio de Janeiro, B. L. Garnier, 1875. xxiii, 359 pp., plate.

805
Catullus, C. V. The Attis of Caius Valerius Catullus; translated into English verse . . . by Grant Allen. London, David Nutt, 1892. xvi, 154 pp.

806 l
— and A. Tibullus. Erotica: the poems of Catullus and Tibullus, and the Vigil of Venus; a literal prose translation with notes, by Walter K. Kelly . . . London, George Bell, 1887. viii, 400 pp. (Bohn's classical library).

807
Cervantes, M. de. Don Quichotte de la manche; traduit de l'Espagnol . . . par Florian. Paris, P. Didot l'Aîné, 1809. 6 vols., plates.

808 a d
— El ingenioso hidalgo Don Quijote de la mancha; nueva edicion, corregida y anotada por Don Eugenio de Ochoa. Paris, Carlos Hingray, 1858. xxxiv, 695 pp., front.

809 d
— The ingenious knight, Don Quixote de la Mancha . . . a new translation . . . by Alexander James Duffield. London, C. Kegan Paul, 1881. 3 vols.

810 k
[Chalesme, — de]. L'homme de qualité, ou les moyens de vivre en homme de bien, & en homme du monde. Paris, André Pralard, 1671. [xii], 254 pp.

811 l
Chapsal, C. P. Modèles de littérature française: ou choix de morceaux en prose et en vers . . . seconde édition. Paris, Alger, L. Hachette, 1848. 2 vols.

812 d
Chasles, P. Etudes sur l'antiquité; précédés d'un essai sur les phases de l'histoire littéraire. Paris, Librairie d'Amyot, 1847. xii, 478 pp.

813 a d h
Chwolson, D. Uber die Uberreste der altbabylonischen Literatur in Arabischen übersetzungen von D. Chwolson. St Petersburg, Buchdruckerei Kaiserlichen Akademie der Wissenschaften, 1859. [ii], 195 pp. [*see* q.].

814
Cicero, M. T. The academic questions, Treatise de finibus, and Tusculan disputations . . . literally translated by C. D. Yonge. London, George Bell, 1878. xxxii, 474 pp. (Bohn's classical library).

815
— On oratory and orators . . . translated or edited by J. S. Watson. London, George Bell, 1876. [vi], 522 pp. (Bohn's classical library).

816
— The orations . . . literally translated by C. D. Yonge. London, George Bell, 1876–8. 4 vols. (Bohn's classical library).

817
— Three books of offices, or moral duties . . . literally translated by Cyrus R. Edmonds. London, George Bell, 1877. viii, 342 pp., front. (port.). (Bohn's classical library).
818
— The treatises . . . on the nature of the gods, on divination, on fate, on the Republic . . . literally translated, chiefly by the editor, C. D. Yonge. London, George Bell, 1878. [vi], 510 pp. (Bohn's classical library).
819 a d h
Clouston, W. A. Arabian poetry for English readers. Glasgow, privately printed, 1881. lxxii, 472 pp., front. illus.
820 d i
— Asiatic and European versions of four of Chaucer's Canterbury tales. [Glasgow], Chaucer Society, [1886]. [iii], 289–436 pp. Part IV of "Chaucer's Analogues". (June 1887).
821 a d
— The book of noodles: stories of simpletons: or, fools and their follies. London, Elliot Stock, 1888. xx, 228 pp.
822 a d h
— A group of Eastern romances and stories from the Persian, Tamil, and Urdu. [Glasgow], privately printed, 1889. xl, 586 pp.
823 a d h i
— Popular tales and fictions, their migrations and transformations. Edinburgh, London, William Blackwood, 1887. 2 vols. (14 Feb. 1887).
824
Corneille, P. Les chef-d'oeuvres dramatiques; nouvelle édition. Tome second à Oxfort, 1770. [iv], 384 pp.
825 e i
Corrêa de Almeida, J. J. Satyras epigrammas e outras poesias. Rio de Janeiro, Eduardo & Henrique Laemmert, 1863. 166 pp. [*see* 253].
826 a d
Cory, I. P. Ancient fragments of the Phoenician, Carthaginian, Babylonian, Egyptian and other authors; a new and enlarged edition . . . by E. Richmond Hodges. London, Reeves & Turner, 1876. xxxvi, 214 pp.
827
Cosson, A. Trozos selectos de literatura y método de composicion literaria sacados de autores arjentinos y estranjeros. Ier tomo. Buenos Aires, Pablo E. Coni, 1869. liv, 367 pp.
828
Costa Rubim, J. F. K. da. Os Inglezes no Brasil: comedia em um prologo e um acto, ao eximio actor brasileiro o Sr. commendador João Caetano dos Santos . . . offerece. Rio de Janeiro, Typ. Portugal e Brazil de J. P. da Silva Rocha, 1863. 26 pp. [*see* 253].
829 k l
Cottin, S. Elisabeth, ou les exilés de Sibérie. Londres, réimprimé par R. Juigné, 1808. iv, 188 pp.
830
Crookenden, I. The silver lock, or Italian banditti: a romance from the manuscript of Isaac Crookenden, of Arundel in Sussex. London, S. Fisher, 1824. 62 pp., front.
831
Dante Alighieri. La divina commedia. Parte prima L'inferno; traduzione ebraica di S. Cav. Dr. Formiggini. Trieste, Tipografia del Lloyd Austriaco, 1869. viii, 205 pp.
832 l
Davidson, J. Ballads and songs; [second edition]. London, John Lane, [etc.], 1894. vi, 131 pp.
833 a d
Defoe, D. The adventures of Captain Singleton. Edinburgh, printed by James Ballantyne, 1810. 2 vols.
834 a
Demosthenes. The Olynthiac and other public orations . . . translated, with notes, &c. by Charles Rann Kennedy. London, George Bell, 1876. 312 pp. (Bohn's classical library).
835
— The orations of Demosthenes against Leptines, Midias, Androtion, and Aristocrates translated, with notes, etc. by Charles Rann Kennedy. London, George Bell, 1877. [v],

407 pp. (Bohn's classical library).

836

— The orations of Demosthenes against Macartatus, Leochares . . . translated by Charles Rann Kennedy. London, George Bell, 1878. [v], 401 pp.

837

— The orations of Demosthenes against Timocrates . . . and for Phormio; translated, with notes and appendices. London, George Bell, 1877. vi, 420 pp. (Bohn's classical library).

838

— The orations of Demosthenes on the crown, and on the embassy; translated with notes, &c. by Charles Rann Kennedy. London, George Bell, 1877. [vi], 401 pp. (Bohn's classical library).

839

— Speeches of Aeschines against Ctesiphon, and Demosthenes on the crown; literally translated into English by Roscoe Mongan. Dublin, William B. Kelly, 1859. 140 pp. (Kelly's classical library).

840 a d

Denis, F. Résumé de l'histoire littéraire du Portugal. Paris, Lecointe et Durey, 1826. xxv, 625 pp.

841 k

[Dering, E. H.]. In the light of the twentieth century, by Innominatus. London, John Hodges, 1886. 153 pp.

842

[Diaz, B.]. Historia de Imperatriz Porcina mulher do Imperador Lodonio de Roma. Rio de Janeiro, Typographia-Popular-de Anzeredo Leite, 1862. 21 pp. [*see* 252].

843 a e

Diogenes Laërtius. The lives and opinions of eminent philosphers; literally translated by C. D. Yonge. London, Henry G. Bohn, 1853. (Bohn's classical library).

844 k

Disraeli, B. Coningsby: or, the new generation. London, Henry Colburn, 1844. 3 vols. Inserted: Key to the characters in Coningsby; comprising about sixty of the principal personages of the story. London, Sherwood, Gilbert & Piper, 1844. 7 pp.

845 d k

— Tancred: or, the new crusade. London, Henry Colburn, 1847. 3 vols.

846 i

[Dowty, A. A. and others]. The Siliad: or, the siege of the seats, by the authors of "The coming K—". London, Ward, Lock & Tyler, [1873]. xv, 236 pp., illus. Lettered on upper cover: . . . Isabel Burton.

847 e

Duthie, W. The pearl of the Rhone and other poems. London, Robert Hardwicke, 1864. viii, 163 pp.

848

Epictetus. The discourses of Epictetus, with the Encheiridion and fragments; translated . . . by George Long. London, George Bell, 1877, xliii, 452 pp. (Bohn's classical library).

849 a d

Ercilla y Zuñiga, A. de and J. de. Vilaviciosa. La Araucana por Don Alonso de y Zuniga y La Mosquea, por D. José de Villaviciosa. Madrid, D. Antonio de San Martin, [etc.]; Barcelona, Librereía de El Plus Ultra, 1861. 2 vols., fronts.

850

Euripides. The tragedies . . . literally translated or revised . . . by Theodore Alois Buckley. London, George Bell, 1877–9. 2 vols., front. (port.). (Bohn's classical library).

851 a

Feuillet, O. La morte; quatre-vingt-deuxième édition. Paris, Calmann Lévy, 1887. [iii], 306 pp.

852 k l

Fouqué, F. H. C. de la Motte. Undine and other tales; translated by F. E. Bunnett. Leipzig, Bernhard Tauchnitz, 1867. [v], 361 pp. Lettered on upper cover: . . . Isabel Burton.

853

Fouquier, A. Chants populaires espangols quatrains et séguidilles. Paris, Librairie des Bibliophiles, 1882. [iii], 146 pp., music 22 pp., plates.

854

[Friswell, J. H.]. A critic critized: a new curiosity of literature. London, Sampson

Low, 1865. 12 pp. Reprinted from the *Examiner,* 11 Feb. 1865 with additions. [*see* 228].

855 a
Gautier, T. Poésies nouvelles. Paris, Charpentier, 1866. [iii], 288 pp.

856
— Premières poésies – 1830–1845. Paris, Charpentier, 1886. [iii], 356 pp.

857 l
Genlis, S. F. de. Jeanne de France; nouvelle historique; nouvelle édition. Paris, Maradan, 1816. 2 vols.

858 a d g
Gesta romanorum: or, entertaining moral stories, invented by the Monks as a fireside recreation . . . translated from the Latin . . . by the Rev. Charles Swan; revised and corrected by Wynnard Hooper. London, George Bell, 1877. lxxvi, 425 pp. (Bohn's classical library).

859 a
Glardon, A. Nouvelles hindoues: Madar; suivi de deux épisodes de la guerre des Cipayes. Lausanne, Georges Bridel, [1885]. 247 pp.

860
Goethe, J. W. Dramatic works of Goethe, comprising Faust, Iphigenia in Tauris . . . translated by Anna Swanwick, and Goetz von Berlichingen. London, George Bell, 1875. xvi, 504 pp. (Bohn's classical library).

861
— Novels and tales, by Goethe, Elective affinities, The sorrows of Werther . . . London, George Bell, 1875. vii, 504 pp. (Bohn's classical library).

862
Goldoni, C. Commedia di tre atti in prosa. In Milano, L. Teofilo Barrois, 1818. 3 pts. (in 1). Contents: L'avventuriere onorato – Pamela – Il vero amico.

863
Goldsmith, O. The miscellaneous works; a new edition; edited by Washington Irving. Paris, W. & W. Galignani, 1825. 4 vols., front.

864 a k
Gonçalves Dias, A. Cantos: collecção de poesias; terceira edicção. Leipzig, F. A. Brockhaus, 1860. xx, 424 pp. front. (port.).

865 k
Goracuchi, G. A. Ecologa per l'anno MDCCCLXXVIII. Trieste, Tipografia del Lloyd Austro-Ung. Editrice, 1878. [ii], 179 pp.

866 a d h
Gosse, Sir E. W. ed. English odes, selected by Edmund W. Gosse. London, C. Kegan Paul, 1881. xxi, 259 pp., front.

867 k
Gowing, E. A. Ballads of the Tower and other poems. London, Griffith Farran Okeden & Welsh, 1891. viii, 152 pp., front.

868 k
— Sita, and other poems, mostly adapted for recitation. London, Elliot Stock, 1895. viii, 104 pp.

869
Greek anthology, The; as selected for the use of Westminster, Eton, and other public schools; literally translated . . . by George Burges . . . London, George Bell, 1876. viii, 519 pp. (Bohn's classical library).

870 a d
Grimm, J. L. C. and W. Grimm. German popular stories and fairy tales, as told by Gammer Grethel . . . revised translation by Edgar Taylor. London, George Bell, 1878. xii, 306 pp., front. (Bohn's illustrated library).

871
Guarini, G. B. Il pastor fido; publicato da A. Buttura. Parigi, Presso Lefevre, 1822. viii, 358 pp., front. (port.). (Biblioteca poetica italiana).

872
Gruen, A. trans. Volkslieder aus Krain. Leipzig, Weidemann, 1850. xxii, 169 pp. [*see* p.].

873 d h
Guest, Lady C. E. The Mabinogion; from the Welsh of the Llyfr coch i hergest . . . in the Library of Jesus College, Oxford; translated, with notes. London, Bernard Quaritch, 1877. xx, 504 pp., illus.

874 a d h

Guest, E. A history of English rhythms; new edition, edited by the Rev. Walter W. Skeat. London, George Bell, 1882. xix, 730 pp.

875 h k

[Gueulette, T. S.]. The thousand and one quarters of an hour (Tartarian tales); edited by Leonard C. Smithers. London, H. S. Nichols, 1893. viii, 308 pp.

876 a d h

Hāfiz, Shīrāzī. English. A specimen of Persian poetry: or odes of Hafez; with an English translation and paraphrase, by John Richardson; new edition . . . by S. Rousseau. London, J. Sewell, [etc.], 1802. xx, 86 pp.

877 a d

— German. Der Diwan des grossen lyrischen Dichters Hafis im Persischen Original; herausgegeben ins deutsch metrisch Ubersetzt . . . von Vincenz Ritter v. Rosenzweig-Schwannau. Wien, K. K. Hof- und Staatsdruckerei, 1858–64. 3 vols.

878 e g j

— Persian. Intikhāb-i dīvān-i Hāfiz [Selections from the dīvān, for the use of schools; edited by Karīm al-Dīn; Persian text. Lahore, 1863]. 96 pp. (litho.). (Edwd. B. Cowell, Oxford, 8 March 1867).

879

— — Ghazaliyāt-i sadr al-'urafā' wa dhuktir al-shu'arā' al-ustād Shams al-Dīn Muhammad Hāfiz [Poems of Hafiz; edited and copied by al-Hāj Muhammad Ibrāhīm; Persian text. Bombay, 1288 A.H., 1871 A.D.]. xxii, 412 pp.

880 g h

— Selections. Háfiz of Shíráz: selections from his poems; translated from the Persian by Herman Bicknell. London, Trübner, 1875. xxi, 384 pp., front., illus., plates. *See also* 105.

881

Harington, Sir J. The most elegant and wittie epigrams. London, George Miller, printer, 1633. [45] pp. [*see* 711].

882 i

Harrison, W. H. The lazy lays, and prose imaginings; written, printed, published, and reviewed by William H. Harrison. London, 1877. 156 pp. (Oct. 1878).

883 i

Hart-Davies, T. L Sind ballads; translated from the Sindi by T. Hart-Davies. Bombay, Education Society's Press, 1881. vi, 47 pp.

884

Hebel, J. P. Schatz-Kästlein rheinischen Hausfreundes. Stuttgart, J. G. Cotta, 1879. viii, 200 pp., illus.

885 a d

Heliodorus, and others. The Greek romances of Heliodorus, Longus, and Achilles Tatius . . . translated from the Greek, with notes, by the Rev. Rowland Smith. London, Henry G. Bohn, 1855. xxxii, 511 pp. At head of title: Scriptores erotici graeci. (Bohn's classical library).

886 a

Herodotus. A new and literal version from the text of Baehr; with a geographical and general index by Henry Cary. London, George Bell, 1877. vi, 613 pp. (Bohn's classical library).

887 a

— Herodoti historiarum. Libri IX; curavit Henr. Rudolph Dietsch [Greek text]. Lipsiae, B. G. Teubner, 1878–9. 2 vols.

888 a d

Hesiod and others. The works of Hesiod, Callimachus, and Theognis; literally translated . . . by the Rev. J. Banks . . . London, George Bell, 1876. xxi, 495 pp. (Bohn's classical library).

889 k

Homer – Iliad – English. The Iliad of Homer; translated by Alexander Pope. London, Henry Lintot, 1743. Vols. 3–6, plates.

890 a d

— — — translated into English . . . by Sir John F. W. Herschel. London, [etc.], Macmillan, 1866. xvii, 549 pp.

891 a d

— — — literally translated with explanatory notes by Theodore Alois Buckley. London, George Bell, 1876. [v], 466 pp. (Bohn's classical library).

892 a d
— — Greek. . . . with copious English notes . . . by the Rev. William Trollope; fifth edition. London, J. & F. H. Rivington, [etc.], 1861. 1, 715 pp.

893 e l
— — — Homeri Ilias . . . nova editio. Lipsiae, Otto Holtze, 1876–7. 2 vols. (in 1).

894
— — Homeri Odyssea; nova editio. Lipsiae, Tauchnitz, 1825. 2 vols. (in 1), front.

895 a d
— — The Odyssey of Homer, with the Hymns, Epigrams, and Battle of the frogs and mice; literally translated with explanatory notes by Theodore Alois Buckley. London, George Bell, 1878. xxxii, 432 pp. (Bohn's classical library).

896
Hooe, W. ed. Authors of the day: or, list of the literary profession for 1879. London, William Poole, 1879. 20 pp.

897 a d
Horace. Q Horatii Flacci Opera Omnia. Oxonii et Londini, Johannes Henricus et Jacobus Parker, 1860. xviii, 270 pp. (Oxford pocket classics).

898 a d
— The works of Horace; translated literally into English prose, by C. Smart; revised . . . by Theodore Alois Buckley. London, George Bell, 1879. viii, 325 pp. (Bohn's classical library).

899 i
Hortis, A. Dante e il Petrarca: nuovi studii. Firenze, 1875. 9 pp. Reprinted from *Rivista europea*. [*see* 986].

900 d i j
— Studj sulle opere latine del Boccaccio. Trieste, Julius Dase, 1879. xx, 956 pp. (Attilio Hortis, Trieste, 15 Nov. 1879). [*see* q.].

901
Howells, W. D. The undiscovered country. Vol. 1. Edinburgh, David Douglas, 1882. 250 pp.

902
Jacobs, F. C. W. Anthologia graeca, sive poetarum graecorum lusus; ex recensione Brunckii . : . Lipsiae, in Bibliopolio Dyckio, 1794–1814. 13 vols.

903 a d h j
Jámí. Persian wit and humour: being the sixth book of the Baháristán of Jámí; translated by C. E. Wilson. London, Chatto & Windus, 1883. vii, 40 pp. (C. E. Wilson, London, 25 April and 2 July 1883).

904 d k
Jeffrey, F. Contributions to the Edinburgh review. London, Longman, Brown, [etc.], 1855. xvi, 1005 pp., front. (port.), plate.

905 a d h
Jeremiah, J. Notes on Shakespeare, and memorials of the Urban club. London, Clayton, 1876. 130 pp., front., (port.), illus., plates.
i
Another copy.

906 h i
— Shakespearean memorabilia: being a collation of all contemporary allusions to the Bard and his works. London, printed for the Editor by C. Skipper & East, 1877. 9 pp.

907 k
Johnson, S. The lives of the most eminent English poets . . . a new edition corrected. London, C. Bathurst, J. Buckland, [etc.], 1783. 4 vols.

908 a d
Justin and others. Justin, Cornelius Nepos, and Eutropius; literally translated . . . by the John Selby Watson. London, George Bell, 1876. xvi, 551 pp. (Bohn's classical library).

909
Juvenal and others. The satires of Juvenal, Persius, Sulpicia, and Lucilius; literally translated . . . by . . . Lewis Evans . . . London, George Bell, 1877. iv, lx, 512 pp., front. (Bohn's classical library).

910 a d g
Kasim ibn 'Alt, called al-Harîri. Makamat: or rhetorical anecdotes of al Harîri of Basra; translated by Theodore Preston. London, James, Madden, [etc.], 1850. xvi, 505 pp.

911 a d
— The assemblies of al Harîri; translated from the Arabic, with an introduction and

notes, by Thomas Chenery. Vol. 1. London, Williams & Norgate, 1870. x, 540 pp.

912

Kathākosha. The Kathákoça: or, treasury of stories; translated from Sanskrit manuscripts by C. H. Tawney, with appendix by Ernst Leumann. London, Royal Asiatic Society, 1895. xxiii, 260 pp.

913

Khalil, al-Khuri. Al-'Asr al-jadid [Poems, second series; Arabic text]. Beirut, 1379 A.H. [1863 A.D.]. 262 pp.

914 k

Kingsford, A. B. Dreams and dream-stories; edited by Edward Maitland; second edition. London, George Redway, 1888. 281 pp.

915

Kirby, W. F. Ed-Dimiryaht, an Oriental romance, and other poems. London, Williams & Norgate, 1867. viii, 238 pp.

916 k

Knowles, J. S. The elocutionist: collection of pieces in prose and verse; twenty-fifth edition; edited by Robert Mullan. Belfast, William Mullan, 1874. xxxvi, 418 pp.

917 d

Koran. Istoria del espanto del dia dez juicio segun las aleyas y profesias del Honrado Alcoran. [Hertford, Imprenta de D. Estevan Austin, 1866]. 16 pp., illus. [*see* p.].

918 a d

Kurz, H. Geschichte der deutschen Literatur mit ausgewählten Stücken aus den Werken der vorzüglichsten Schriftsteller. Leipzig, B. G. Teubner, 1853–9. 3 vols.

919

La Fontaine, J. de. Contes et nouvelles; nouvelle édition. Paris, Garnier Frères, [1861?]. viii, 443 pp., front.

920

Leroux, P. J. Dictionnaire comique, satyrique, critique, burlesque . . . nouvelle édition. A Pampelune, 1786. 2 vols.

921 k

Lever, S. Fireflies: ballads and verses. London, Remington, 1883. xi, 133 pp.

922

Livingstone, W. P. The Negro Prince: or, the victims of Dahomey. London, Charles Henry Clarke, 1862. iv, 331 pp.

923 a d l

Livy. The history of Rome; the first eight books, literally translated with notes and illustrations, by D. Spillan. London, Bell & Daldy, 1872–3. 4 vols. (Bohn's classical library).

924 a d

Lotti, L. La liberazione di Vienna assediata dalle armi ottomane, poemetto giocoso e la Banzuola, dialoghi sei . . . in lingua popolare bolognese. [n.p., 1690?]. [viii], 248 pp., front., illus.

925 a d

[Louis XI, Roi de France, 1423–83]. Les cents nouvelles nouvelles; texte revu avec beaucoup de soin sur les meilleures éditions. Paris, Garnier Frères, [1865]. xxix, 424 pp.

926 d g

Lucan. The Pharsalia; literally translated into English prose with copious notes by H. T. Riley. London, George Bell, 1878. xi, 427 pp. (Bohn's classical library).

927 d l

Lucian. Walker's Selections from Lucian, with copious English notes, and a much improved lexicon by Henry Edwardes. London, William Allan, 1859. [iv], 110 pp.

928 a d

Lucretius. Lucretius on the nature of things: a philosophical poem in six books; literally translated into English prose by John Selby Watson . . . London, George Bell, 1876. xxiii, 496 pp. (Bohn's classical library).

929 a d h

Lyall, C. J. Translations of ancient Arabian poetry, chiefly prae-Islamic, with an introduction and notes. London, Williams & Norgate, 1885. lii, 142 pp.

930 a l

MacCall, W. Russian rhymes: London, George Standring, 1878. 63 pp. Date on wrapper 1879. [*see* p.].

931 i
McCarthy, J. H. Serapion and other poems. London, Chatto & Windus, 1883. viii, 177 pp.

932 i j
— Hafiz in London. London, Chatto & Windus, 1886. viii, 90 pp. (Justin Huntly McCarthy, London, 3 May 1886).

933
MacDowall, C. The first empress of the East. London, W. H. Beer, 1886. [iii], 106 pp.

934 i l
— Lady Margaret's sorrows . . . and other poems. London, W. H. Beer, 1883. 120 pp.

935 k
Maffei, G. Storia della letteratura italiana; terza edizione. Volume secondo. Firenze, Felice Le Monnier, 1853. [i], 490 pp.

936 d k
Mahābhārata. Indian idylls from the Sanskrit of the Mahâbhârata by Edwin Arnold. London, Trübner, 1883. xiii, 282 pp. (Trübner's oriental series).

937 k
Main, D. M. ed. Three hundred English sonnets. Edinburgh, London, William Blackwood, 1884. [viii], 320 pp.

938 a d h
— ed. A treasury of English sonnets; edited from the original sources. Manchester, Alexander Ireland, printer, 1880. viii, 471 pp.

939
Mann, A. M. Shadow and sunlight: poems (Johannesburg) 1890–1893. Johannesburg, J. C. Juta, 1893. x, 265 pp.

940 h i
[Marchesetti, C.]. A muzio de Tommasini pell'ottantesimo anniversario carme. [Vienna], Stab. Tip. Appolonio & Caprin Edit., [1874]. 12 pp. [*see* p.].

941 a d
Margaret d'Angoulême, Queen Consort of Henry II, King of Navarre. L'Heptaméron: contes de la Reine de Navarre; nouvelle édition. Paris, Garnier, [1877?]. xii, 468 pp.

942
Martial. The epigrams of Martial; translated into English prose, each accompanied by one or more verse translations . . . London, George Bell, 1877. x, 660 pp. (Bohn's classical library).

943 d l
— M. Val. Martialis ex Museo Petri Scriverii. Amstelodami, apud Ioann. Linssonium, [*c*. 1628]. 263 pp.

944 d
Mārun Naqqāsh. Arzah lubnān [Four comedies, to which are added some poems and fragments; edited, with an introduction and a life of the author by Niqūlā Naqqāsh; Arabic text]. Beirut, 1869. 496 pp.

945 a d h
Massey, G. The secret drama of Shakespeare's sonnets unfolded; second and enlarged edition . . . for subscribers only. London, Spottiswoode, printer, 1872. xii, 603, 56 pp.

946 i
Mehmed beg Kapetanović Ljubušak. Narodno blago sakupio i izdao Mehmed beg Kapetanović Ljubušak po Bosni, Hercegovini i susjednim krajevima. Sarajevo, Troškom vlastnika, 1887. 463 pp.

947 i
Metastasio, P. Alcune lettere inedite; pubblicate dagli autografi da Attilio Hortis. Trieste, Tipografia del Lloyd Austro-Ungarico, 1876. lii, 100 pp. (Attilio Hortis).

948
Métivier, G. Poësies guernesiaises et françaises avec glossaire. Guernesey, Thomas-Mauger Bichard, 1883. xiii, 324, xlvii pp.

949
Millington, J. intro. English as she is spoke: or a jest in sober earnest. London, Field & Tuer, [etc. 1883]. [4], xv, 60 pp.

950 k
Milnes, R. M., Baron Houghton. Palm leaves. London, Edward Moxon, 1844. xxxvi, 202 pp.

951 a d h
Milton, J. The sonnets; edited by Mark Pattison. London, Kegan Paul, Trench, 1883. 227 pp., front.

952 a
Moeckeschl, M. S., trans. Haideblümchen: zigeunerische Dichtungen und Sprichwörter. Bukarest, Thiel & Weiss, 1873. 58 pp. [*see* p.].
953 a d
Montesquieu, C. de S., Baron de. Lettres persanes suivies de Arsace et Isménie et de pensées diverses; nouvelle édition. Paris, Garnier Frères, [1866]. [iii], 431 pp.
954 d
Morley, H. Of English literature in the reign of Victoria, with a glance at the past. Leipzig, Bernhard Tauchnitz, 1881. xl, xii, 416 pp.
955 a
Mu'allakat. The Moallakát: or seven Arabian poems, which were suspended on the temple at Mecca; with a translation, and arguments by William Jones. London, J. Nichols, printer for P. Elmsley, 1783. [iv], 163 pp., front. (port.), tables.
956 k
Mudie, C. E. Stray leaves; a new edition. London, Macmillan, 1873. 82 pp.
957 a d
Mueller, F. M. A history of ancient Sanskrit literature so far as it illustrates the primitive religion of the Brahmans. London, Williams & Norgate, 1859. xix, 607 pp.
958 d k
Naden, C. C. W. Songs and sonnets of springtime. London, C. Kegan Paul, 1881. xi, 171 pp.
959
Nasr al-Dīn, Khojah. Mashūr Khwājah Nasr al-Dīn latā'ifi [Jests of Nasr al-Dīn Khojah; Turkish text]. 40 pp., illus. [*see* p.].
960 d
— Nawādir al-Khwājah Nasr al-Dīn, juhā al-Rūm [Anecdotes of Khojah Nasr al-Dīn, the Turkish jester; translated from the Turkish into Arabic]. Cairo, 1280 A.H. [1864 A.D.]. 60 pp. [*see* p.].
961 a d
— Les plaisanteries de Nasr-Eddin Hodja; traduites du Turc par J.-A. Decourdemanche. Paris, Ernest Leroux, 1876. 108 pp.
962
Nepos, C. Liber de excellentibus ducibus exterarum gentium cum vitis Catonis et Attici. Edinburgh, William & Robert Chambers, 1852. xi, 210 pp. [*see* 642].
963
[Newman, J. H., Cardinal]. Dream of Gerontius; twenty-eighth edition. London, Longmans, Green, 1892. 60 pp.
964 a d
Nùnzi, C. Poesí in dialètt bulgnèis. Bulògna, Stamparí Militar, 1874. 96 pp.
i
Two other copies.
965 a
Ohnet, G. Le docteur Rameau; cent deuxième édition. Paris, Paul Ollendorff, 1889. 345 pp. (Les batailles de la vie).
966
— Le maître de forges; deux cent septième édition. Paris, Paul Ollendorff, 1885. [iv], 403 pp. (Les batailles de la vie).
967 a k
[Oliphant, M.]. Two stories of the seen and the unseen. Edinburgh, London, William Blackwood, 1885. [iii], 212 pp.
968
Oliveira Menezes R. O. de. Haabás drama em um prologo e dous actos. [São Paulo, Typ. Litteraria, 1861]. iv, 55 pp. [*see* 249].
969
Omar Khayyám. Ruba'iyyāt-i Hakīm Khayyām [followed by quatrains of Bābā Tāhir Hamadānī, 'Attār, and Irich Mīrzā (Insāf), a tarji' by Shams-i Tabrīzī, and one by Nāsir-i Khusrau; Persian text]. [Persia], 1284 A.H. [1867 A.D.]. [102] pp. (lithograph) [*see* p.].
970 a d h
— Rubáiat of Omar Khayyám . . . rendered into English verse; third edition. London, Bernard Quaritch, 1872. xxiv, 36 pp.
971 d
— — and the Salámán and Absál of Jámí; rendered into English verse; [fourth edition]. London, Bernard Quaritch, 1879. [4], xvi, 112 pp., front.

972 a d h i
— — translated by Justin Huntly McCarthy. London, David Nutt, 1889. lxii, clvii pp.
973
[Ord, J. W. and J. R. Robinson. Poems. Stokesley, 1872]. 265–96 pp., port. From *The bards and authors of Cleveland and south Durham* [etc.] by G. M. Tweddell. [*see* 251].
974 k
Ossian. The poems of Ossian . . . translated by James Macpherson; Imray's edition. Vol. II. Glasgow, Chapman & Lang, printer for J. Imray, 1799. [i], 236 pp., front., plates.
975 d
Ovid. The Fasti, Tristia, Pontic epistles, Ibis, and Halieuticon; literally translated . . . by Henry T. Riley. London, George Bell, 1878. xxiv, 503 pp. (Bohn's classical library).
976 d
— The metamorphoses; literally translated . . . by Henry T. Riley. London, George Bell, 1877. xiv, 554 pp. (Bohn's classical library).
977
Passow, A. ed. Popŭlaria carmina graeciae recentioris. Lipsiae, B. G. Teubner, 1860. xi, 650 pp.
978
Pausanias. Description of Greece; translated . . . by Arthur Richard Shilleto. London, George Bell, 1886. 2 vols. (Bohn's classical library).
979 a
Payne, J. Intaglios sonnets; new edition. London, W. H. Allen, 1884. vi, 82 pp.
980
— Lautrec: a poem; new edition. London, W. H. Allen, 1884. 59 pp.
981
— The masque of shadows and other poems; new edition. London, W. H. Allen, 1884. [xiii], 222 pp.
982 i
— New poems; new edition. London, W. H. Allen, 1884. viii, 295 pp.
983
— Songs of life and death; new edition. London, W. H. Allen, 1884. [2], xvi, 262 pp.
984
Pereira de Fonseca, M. J., Marquis de Maricá. Colleccão completa das maximas pensamentos e reflexões; ediçao revista. Rio de Janeiro, Eduardo e Henrique Laemmert, 1843. viii, 376 pp.
985
Petrarch. Le rime . . . publicate da A. Buttura. Parigi, Lefevre, 1820. 3 vols.
986 h i
— Scritti inediti . . . pubblicati ed illustrati da Attilio Hortis. Trieste, Tipografia del Lloyd Austro-ungarico, 1874. xvi, 372 pp., illus.
987 a d
— The sonnets, triumphs and other poems . . . translated . . . by various hands. London, George Bell, 1875. cxl, 416 pp., front. (port.), plates. (Bohn's illustrated library).
988 a d
Pfeiffer, E. Poems; second edition. London, C. Kegan Paul, 1878. 150 pp., illus.
989 d
Phaedrus. Fabulae selectae. London, Edinburgh, William & Robert Chambers, 1853. xi, 112 pp. (Chambers's educational course, classical section). [*see* 642].
990 i
[Pickering, C. J.]. The last David, and other poems. London, Elliot Stock, 1883. [iii], 123 pp.
991 a i j
— Metassai: scripts and transcripts. Glasgow, privately printed, 1887. 156 pp. (Chas. J. Pickering, Bath, 9 Feb. 1887).
992
— 'Umar of Nîshâpûr. [London], 1890. 16 pp. Reprinted from the *National review*, Vol. 16, No. 94. [*see* p.].
993 a d
Pigott, G. A manual of Scandinavian mythology. London, William Pickering, 1839. xliv, 370 pp.
994 i
Pincherle, J. Il cantico dei cantici di Salomone, per la prima volta tradotto dal testo

ilaniano in fronte nell'idioma zingaresco (Indo-orientale): studio. Trieste, Giovanni Balestra, printer, 1875. [vi], 14 pp. [*see* p.].

995 l

— Il viaggio sentimentale di Sterne continuato. Trieste, Tipografia Weis, 1871. [iii], 67 pp. [*see* p.].

996 a d

Pindar. The odes of Pindar; literally translated . . . by Dawson W. Turner. London, George Bell, 1876. xxvii, 434 pp., front. (port.). (Bohn's classical library).

997

Pitts, J. L. The patois of the Channel islands; the Norman-French text, edited with parallel English translation. Guernsey, Guille-Allès Library, MacKenzie & Le Patourel, 1883. viii, 62 pp., front., illus. [*see* p.].

998 d

Plato. Summary and analysis of The dialogues . . . by Alfred Day. London, Bell & Daldy, 1870. viii, 530 pp. (Bohn's classical library).

999 a d

— The works of Plato; a new and literal version . . . by Henry Cary [and others]. London, George Bell, 1876–7. 6 vols., front. (port.).

1000 a d

Plautus. The comedies of Plautus literally translated . . . by Henry Thomas Riley. London, Henry G. Bohn, George Bell, 1852. 2 vols.

1001 a d

Pliny. The natural history of Pliny; translated . . . by John Bostock and H. T. Riley. London, Henry G. Bohn, 1855–7. 6 vols.

1002 k

[Poillow, G. A., Vicomtesse de Saint-Mars]. Les galanteries de la cour de Louis XV, par La Comtesse Dash; nouvelle édition. Paris, Michel Lévy, 1867. [ii], 310 pp.

1003 a d

Propertius. The elegies of Propertius . . . literally translated by R. Brinsley Sheridan and Mr. Halhed; edited by Walter K. Kelly. London, George Bell, 1878. xi, 500 pp. (Bohn's classical library).

1004 a

Purānas. Skandapurāna. The Sahyâdri-khanda of the Skanda Purâna: a mythological, historical, and geographical account of western India; first edition of the Sanskrit text . . . by J. Gerson da Cunha. Bombay, Thacker, Vining, [etc.], 1877. [iii], [576] pp.

Another copy.

1005 d i

Quental, A. de. Ausgewählte Sonette aus dem Portugiesischen verdeutscht von Wilhelm Storck. Paderborn, Münster, Ferdinand Schöningh, 1887. 126 pp.

1006 e i

— Os sonetos completos; publicados por J. P. Oliveira Martins. Porto, Lopes, 1886. 48, 126 pp.

1007

Quintilian. Institutes of oratory or, education of an orator . . . literally translated . . . by John Selby Watson. London, George Bell, 1857–6. 2 vols. (Bohn's classical library).

1008 d

Rabelais, F. Oeuvres. Paris, Louis Janet, 1823. 3 vols.

1009 k

Ramsden, Lady G. Speedwell. London, Richard Bentley, 1894. [v], 291 pp.

1010 i

[Reeve, A.]. Faithful for ever: a poem, by A.R. London, Robert Banks, 1882. 105 pp.

1011 a h

— Lights and shadows: poems. London, S. W. Partridge, 1883. iv, 248 pp.

1012 k

Regan, T. Love's pilgrim [and other pieces. London, *c.* 1891]. 13 pp., 6 sheets of music. [*see* pq.].

1013 a d

Richards, A. B. Croesus, king of Lydia: a tragedy in five acts; second edition, revised. London, Longman, Green, [etc.], 1861. xvi, 113 pp.

1014

— Medea: a poem. London, Chapman & Hall, 1869. x, 67 pp.

1015

— Oliver Cromwell: an historical tragedy, in a prologue and four acts; third edition.

London, Effingham Wilson, 1873. 73 pp.

1016
Robida, A. Le vingtième siècle, texte et dessins. Paris, [F. Aureau, Imprimerie de Lagny], 1883. 425 pp., illus., plates.

1017 a d
Roebuck, T. A collection of proverbs, and proverbial phrases, in the Persian and Hindoostanee languages; compiled and translated. Calcutta, Hindoostanee Press, 1824. 2 pts. (in 1).

1018 d i
Rolland, J. Ane treatise callit The court of Venus . . . newlie compylit be Iohne Rolland in Dalkeith, 1575; edited by Walter Gregor. Edinburgh, London, William Blackwood for the Scottish Text Society, 1884. xxxii, 231 pp.

1019 a d
Rousseau, J. J. Les confessions . . . suivies des rêveries du promeneur solitaire. Lausanne, François Grasset, 1782. 2 vols.

1020 a d
— Emile: ou de l'éducation; nouvelle édition. Paris, Garnier Frères, [1887]. [iii], 567 pp.

1021
Rufus, Q. C. De gestis Alexandri magni, Regis Macedonum, libri qui supersunt VIII. Edinburgh, William & Robert Chambers, 1849. 352 pp., front. (map), illus. (Chambers's educational course, classical section). [*see* 642].

1022 a d
Sacy, S. de. Chrestomathie arabe: ou extraits de divers écrivains arabes, tant en prose qu'en vers, avec une traduction française et des notes; seconde édition. Paris, Imprimerie Royale, 1826-7. 3 vols.

1023 d l
Sa'dī, Shīrāzī. The Persian and Arabick works of Sâdee; [edited by John Herbert Harington]. Calcutta, The Honorable Company's Press, printer, 1791-5. 2 vols. [*see* q.].

1024
Sallust, C. C. Catilinaria et Jugurthina bella; editio stereotypa. Parisiis, P. et F. Didot, 1801. 150 pp.

1025 a d
— and others. Sallust, Florus, and Velleius Paterculus, literally translated by John Selby Watson. London, George Bell, 1876. xvi, 560 pp. (Bohn's classical library).

1026 d
Santa Rita Durão, J. de. Caramurú: poema epico do descubrimento da Bahia; segunda edição. Lisboa, Imprensa Nacional, 1836. 307 pp.

1027
Selenidi, B. Il mondo della luna: poema fantastico-umoristico. Trieste, G. Balestra, 1876. 15 pp. [*see* p.].

1028 d h j
Şeyhzade. The history of the forty vezirs: or the story of the forty morns and eves, written in Turkish by Sheykh-Zāda; done into English by E. J. W. Gibb. London, George Redway, 1886. xl, 420 pp. (E. J. W. Gibb, Glasgow, 26 Aug. and 6 Sept. 1886).

1029
Shakespeare, W. Hamlet: drama em cinco actos; traducçao portugueza; segunda ediçao. Lisboa, Imprensa Nacional, 1880. 149 pp.

1030 a l
— Hamlet, Prince of Denmark; lines pronounced corrupt and restored . . . with preface and notes . . . by Matthias Mull. London, Kegan Paul, Trench, 1885. lx, 120 pp.

1031 i
— A new variorum edition of Shakespeare; edited by Horace Howard Furness. Vol. VII: The merchant of Venice. Philadelphia, J. B. Lippincott, 1888. xii, 479 pp. (H. H. Furness).

1032 a
Shelley, P. B. Der entfesselte Prometheus: lyrisches Drama in vier Akten; Deutsch von Albrecht Graf Wickenburg. Wien, L. Rosner, 1876. xxiii, 103 pp.

1033
Signorini, A. trans. Cento racconti per bambini intorno agli animali, compilati da A.J.P.; tradotti dall'inglese. Firenze, Tipografia Cooperativa, 1885. 95 pp., front., plates.

1034
Simrock, K., trans. Das Nibelungenlied; Schulausgabe, mit Einleitung und Wörterbuch. Stuttgart, J. G. Cotta, 1874. xii, 310 pp.
1035
— — vierzigste Auflage. Stuttgart, J. G. Cotta, 1880. vii, 387 pp., front.
1036
Sirovich, A. I martiri della Serbia: racconto storico del secolo XIX. Trieste, G. Balestra, 1876. [vii], 483 pp.
1037
Slovénşke pésmi, Krajnfkiga naróda, zheterti svésik. V Ljubljani, Joshef Blasnik, 1841. 144 pp.
1038 a
Smith, L. A. Through Romany songland. London, David Stott, 1889. xix, 227 pp.
1039 i
Smollett, T. The works of Tobias Smollett; carefully selected and edited . . . by David Herbert. Edinburgh, William P. Nimmo, 1871. 623 pp., front. (port.).
1040
Somma, M. ed. Cento racconti . . . in questa terza edizione. Napoli, Francesco Migliaccio, 1822. 328 pp. [*see* 743].
1041
Sophocles. The tragedies of Sophocles; the Oxford translation; new edition. London, George Bell, 1878. xvi, 339 pp., front. (port.).
1042 a d
Somadeva Bhatta. The Kathá sarit ságara: or ocean of the streams of story; translated from the original Sanskrit, by C. H. Tawney. Calcutta, Baptist Mission Press for Asiatic Society of Bengal, 1880–7. 2 vols. (Biblioteca indica).
d
Another copy of Vol. 2, fasc. XI and XIII.
1043
Souza Menezes, R. I. de. O solitario: romancete. Bahia, Typ. de Camillo de Lellis Masson, 1857. [ii], 93 pp. [*see* 246].
1044 d
Spitta, G. compiler. Contes arabes modernes; recueillis et traduits. Leide, E. J. Brill; Paris, Maisonneuve, 1883. xi, 225 pp.
1045 l
Stampella, A. Un altro Misogallo rime. Trieste, Levi, 1871. 104 pp. [*see* p.].
1046
Stanley, H. E. J. The poetry of Mohammed Rabadan. [London, 1870]. 61–100 pp. Reprinted from the *Journal of the Royal Asiatic Society,* N.S. Vol. 4, pp. 138–77. [*see* 257].
1047 a d
— ed. Rouman anthology: or selections of Rouman poetry, ancient and modern, being a collection of the national ballads of Moldavia and Wallachia. Hertford, Stephen Austin, 1856. xx, 226 pp.
1048 k
Sterne, L. The beauties of Sterne; including all his pathetic tales, & most distinguished observations on life . . . eighth edition. London, G. Kearsley, 1785. xviii, 197 pp.
1049 a
[Stoddart, J. H.]. The village life. Glasgow, James Maclehose, 1879. 206 pp.
1050 i
Storck, W. Buch der Lieder aus der Minnezeit. Münster, Adolph Russell, 1872. xxvi, 399 pp.
1051 i
— Lose Ranken: ein Büchlein catallischer Lieder. Münster, E. C. Brunn, 1867. 142 pp. (Münster, 1 Dec. 1883).
1052 a
— trans. Hundert altportugieschische Lieder; zum ersten Male Deutsch. Paderborn, Münster, Ferdinand Schöningh, 1885. vii, 124 pp. [*see* p.].
1053
Storm, T. Hausbuch aus deutschen Dichtern seit Claudius: eine kritische Anthropologie; erste illustrirte Ausgabe . . . von Hans Speckter. Leipzig, Wilhelm Mauke, 1875. [viii], 462 pp.
1054 d
Swinburne, A. C. Essays and studies. London, Chatto & Windus, 1875. xiii, 380 pp.

1055
— Ode on the proclamation of the French Republic, September 4th, 1870. London, F. S. Ellis, 1870. 23 pp. [*see* 222].

1056 e i l
Sylvester, J. J. The laws of verse: or principles of versification. London, Longmans, Green, 1870. 152 pp.

1057 a d
Tacitus. The works of Tacitus; the Oxford translation, revised. London, George Bell, 1877. 2 vols. (Bohn's classical library).

1058
Taine's History of English literature; [review article. London, 187-]. 473-512 pp. From an unidentified journal. [*see* 225].

1059 a
Talbot de Malahide, J. Baron. The light literature of Spain, particularly the works of Fernan Caballero. Dublin, Hodges, Foster, 1872. 32 pp [*see* p.].

1060
Tansillo, L. The nurse: a poem; translated from the Italian . . . by William Roscoe; second edition. Liverpool, Cadell & Davies, 1800. 31, 89, 34 pp., illus.

1061
Tasso, T. Aminta favola boschereccia; publicata da A. Buttura. Parigi, Lefèvre, 1822. [iii], 121 pp., front.

1062 a d h
— Godfrey of Bulloigne: or, the recovery of Jerusalem . . . from the Italian . . . by Edward Fairfax; the fifth edition . . . Windsor, R. S. Kirby, 1817. 2 vols.

1063 d
— The Jerusalem . . . translated into English verse, by John Kingston James. London, Longman, Green, [etc.], 1865. 2 vols.

1064 a d
Terence and Phaedrus. The comedies of Terence, and the fables of Phaedrus; literally translated . . . by Henry Thomas Riley. London, George Bell, 1879. viii, 535 pp., front. (port.).

1065 a d
Theocritus and others. The idylls of Theocritus, Bion, and Moschus . . . translated by J. Banks. London, George Bell, 1876. xxiv, 343 pp., front. (port.). (Bohn's classical library).

1066 k
Thompson, F. Poems. London, Elkin Mathews & John Lane, [etc.], 1894. viii, 81 pp. (Constance Wilde, Jan. 1894).

1067
Thucydides. Analysis and summary of Thucydides . . . [by J. T. Wheeler]. London, George Bell, 1876. xvi, 413 pp. (Bohn's collegiate series, philological library).

1068
— The history of the Peloponnesian war; a new and literal version, by Henry Dale. London, George Bell, 1878. 2 vols. (Bohn's classical library).

1069 i
Trevors, T. Fugitive poems. Dover, "Dover Telegraph" Office, printer, 1861. 48 pp.

1070
Turner, D. W. Notes on Herodotus, original and selected from the best commentators; second edition, with numerous additions . . . London, George Bell, 1876. xxiv, 472 pp. (Bohn's philological library).

1071
Twain, M. pseud. A curious dream, and other sketches; selected and revised by the author. London, George Routledge, [1872]. 150 pp.

1072 a d
Valderrama, A. Bosquejo histórico de la poesía chilena. Santiago, Imprenta Chilena, 1866. 270 pp.

1073 d
[Varnhagen, F. A. de]. Epicos brasileiros; nova edicçâo. [Lisboa, Imprensa Nacional], 1845. 453 pp.

1074
Vassalli, M. Motti, aforismi e proverbii maltesi. Malta, Stampato per l'autore, 1828. vii, 93 pp.

1075
Vay, A. von. Erzählungen des ewigen Mütterleins. Budapest, Ferdinand Tettey, 1879. 302 pp.

1076 d
Veiga, L. F. da. intro. Cartas chilenas (treze) . . . copiadas de um antigo manuscripto; com uma introduccão. Rio de Janeiro, Eduardo & Henrique Laemmert, 1863. 223 pp.
1077 d
Venizelos, I. Paroimiai dēmōdeis sullegeisai kai 'ermēneutheisai; ekdosis deutera epēuxsēmenē kai diorthōmenē [Popular proverbs collected and explained; second edition, enlarged and corrected; Greek text]. En 'Ermoupolei, ek tou tupografeiou tēs "Patridos", 1867. [viii], 360 pp. [*see* 706].
1078 i
Viale, A. J. Bosquejo metrico da historia de Portugal. Lisboa, Imprensa Nacional, 1866. [iii], 263 pp.
1079 a d
Vikramāditya, King of Ujjayiní. The Buttris Singhasun: or, the tales of the thirty-two images, being a series of amusing anecdotes of the celebrated Hindoo Prince Sree Vicrumaditiu; translated into English (for the first time) from the original Bengali, by Cheedam Chunder Das. Calcutta, 1817. xi, 209 pp.
1080 l
— Singhāsan battīsī: or, the thirty-two tales of Bikramājīt; translated into Hindī from the Sanskrit, by Lallūjī Lāl Kabi; a new edition revised by Syed Abdoollah. London, Wm. H. Allen, 1869. xvi, 216 pp.
1081
— Sing, hasun Butteesee, or anecdotes of the celebrated Bikrmajeet; translated into Hindoostanee from the Brij-B, hak, ha of Soondur Kubeeshwur, by Meerza Kazim Ulee Juwan, and Shree Lulloo Lal Kub . . . [in Hindi]. Calcutta, Hindoostanee Press, printer, 1805. 252 pp. [*see* q.].
1082 a d h
Villon, F. The poems . . . now first done into English verse . . . by John Payne. London, Reeves & Turner, 1881. xcvi, 150 pp.

1083 i
Virgil. Le georgiche; tradotte in ottava rima da Francesco Combi. Venezia, Tipografia Antonelli, 1873. xxiv, 115 pp. (Francesco Combi).
1084 a d
— Opera omnia cum notis selectissimis variorum, Servii, Donati, Pontani, Farnabii, &c. et indice . . . Cornelii Schrevelii. Lugd. Batavorum, apud Franciscum Hackium, 1652. [xvi], 1051 pp.
1085 k
— The works . . . in English verse; translated by Christopher Pitt . . . [and] by Joseph Warton. London, R. & J. Dodsley, 1763. 4 vols., fronts, plates, map.
1086 a l
— The works; literally translated . . . by Davidson; a new edition . . . by Theodore Alois Buckley. London, George Bell, 1877. ix, 404 pp., front. (port.) (Bohn's classical library).
1087 a d
Walker, W. S. ed. Corpus poetarum latinorum; edidit Gulielmus Sidney Walker. Londini, Henry G. Bohn, 1854. vi, 1209 pp.
1088 l
[Warner, R.]. Bath characters: or sketches from life, by Peter Paul Pallet; second edition. London, G. Wilkie & J. Robinson, 1808. xxiv, 132 pp.
1089
Wheeler, J. T. An analysis and summary of Herodotus; with a synchronistical table of principal events; second edition, revised . . . London, George Bell, 1877. xvi, 349 pp. (Bohn's philological library).
1090 a
— A dictionary of the noted names of fiction; a new edition. London, George Bell, 1876. xxxiv, 410 pp.
1091
Wickenburg-Almásy, W., Gräfin. Seeröslein: ein Märchen in Versen. Wien, K. K. Hof- und Staatsdruckerei, 1877. 56 pp. [*see* p.].
1092
Wilding, E. Songs of passion and pain. London, Newman, 1881. iv, 100 pp.
1093 a d
Wilson, H. H. [Select specimens of the theatre of the Hindus]. Calcutta, V. Holcroft, Asiatic Press, 1826. 4 vols. (in 1). Contents: 1, The Mrichchakati, or the toy cart –

2, Vikrama and Urvasi, or the hero and the nymph – 3, Malati and Madhava, or the stolen marriage – 4, Uttari Rama Cheritra, or continuation of the history of Rama.

1094 a d
Wolf, F. Le Brésil littéraire: histoire de la littérature brésilienne. Berlin, A. Asher, 1863. 2 pts. (in 1).

1095 a d
Xenophon. The Anabasis: or expedition of Cyrus, and the Memorabilia of Socrates, literally translated from the Greek, by J. S. Watson. London, George Bell, 1878. viii, 519 pp., front. (port.).

1096 a d
— The Cyropaedia: or institution of Cyrus, and the Hellenics: or Grecian history; literally translated from the Greek, by J. S. Watson . . . and Henry Dale. London, George Bell, 1876. xvi, 580 pp. (Bohn's classical library).

1097
— Minor works; literally translated from the Greek, with notes and illustrations, by J. S. Watson. London, George Bell, 1878. [vii], 384 pp. (Bohn's classical library).

1098 l
— Scripta quae supersunt, Graece et Latine. Parisiis, Ambrosio Firmin Didot, 1860. xxiv, 799 pp.

1099
Yahyā ibn Sharaf (Abu Zakarīyā), al-Nawawī. Kitāb al-majālis al-sanīyah fi al-Kalām 'alā al-arba'in al-Nawawīyah, al-Arba'ūn hadīthan [A collection of forty select traditions, accompanied by a diffuse commentary entitled al-majālis al-saniyah, by Ahmad ibn Hijāzī al-Fashnī; Arabic text]. Cairo, 1285 A.H. [1868 A.D.]. 251 pp.

1100 a i j
Youngs, E. S. The apotheosis of Antinoüs and other poems. London, Kegan Paul, Trench, 1887. viii, 213 pp., front. (port.).

1101 k
— A heart's life: Sarpedon and other poems. London, Kegan Paul, Trench, 1884. viii, 166 pp.

1102
Zimmermann, J. G. Solitude . . . to which are added, the life of the author . . . London, Vernon & Hood, [etc.], 1978. lii, 330 pp., front., illus., plates.

1103 a d
Zorutt, P. Poesiis. Udin, Stamperie Vendram, 1846–7. 2 vols. (in 1).

1104
— Poesiis. Udin, Stamperie Murer, [1847–57]. 2 vols. (in 1).

1105 a
Zotenberg, J. Notice sur le livre de Barlaam et Joasaph, accompagnée d'extraits du texte grec et des versions arabe et éthiopienne. Paris, Imprimerie Nationale, 1886. [v], 166 pp. Reprinted from *Notices et extraits des manuscrits de la Bibliothèque nationale,* Vol. 28.

1106–81 Medicine, Psychology

1106
Anonymous. Arnica montana and rhus toxicodendron as external remedies . . . new edition. London, James Leath, 1856. x, 78 pp.

1107
— The family physician: a manual of domestic medicine, by physicians and surgeons of the principal London hospitals. London, [etc.], Cassell Petter & Galpin, [1879]. xxxii, 1022 pp., illus.

1108
— London medical practice: its sins and shortcomings, by a physician [*i.e.* S. Dickson?]. London, Simpkin, Marshall, 1860. [ii], 81 pp. [*see* 224].

1109
— "Miraculous cures": can such phenomena be accounted for upon natural principles? [London, 18—]. 12 pp. [*see* p.].

1110 l
Achillinus, A. De chyromantiae principiis et physionomiae [and] Bartholomei Coclitis Chyromantie ac physionomie anastasis; cuz approbatiõe magistri Alexãdri ₫ Achillinis. [Bononiae, Ioanez Antoniuz Platonidem Benedictor uz, 1503]. [24, 328] pp., illus. [*see* q.].

1111 a d h i j
Allen, E. H. A manual of cheirosophy: being a complete practical handbook of the twin sciences of cheirognomy and cheiromancy . . . London, New York, Ward, Lock, 1885. 319 pp., illus. (Ed. Heron Allen, 18 Oct. 1885; 27 Feb. 1886).

1112 e h i
Arpentigny, C. S. d'. La science de la main – The science of the hand: or, the art of recognising the tendencies of the human mind by the observation of the formations of the hands; translated from the French . . . , and edited with an introduction . . . by Ed. Heron-Allen. London, New York, Ward, Lock, 1886. 444 pp., front., illus., plates, bibliogr. (Ed. Heron Allen).

1113 e
Baughan, R. The handbook of palmistry. London, George Redway, [1885]. 32 pp. [*see* p.].

1114 k
Baynes, D. Auxiliary methods of cure: the Weir Mitchell system, massage, Ling's Swedish movements, the hot water cure, electricity. London, Simpkin, Marshall, 1888. 138 pp.

1115 d
[Bérard, S.]. Count Matteps electro-homoeopathic specifics: a new science; translated from the French. London, Leath & Ross, 1880. xii, 281 pp., front. (port.), plates, tables.

1116
[Berdoe, E.]. Dying scientifically: a key to St. Bernard's, by Aesculapius Scalpel. London, Swan Sonnenschein, 1888. 120 pp.

1117
— St. Bernard's: the romance of a medical student, by Aesculapius Scalpel; second edition, revised. London, Swan Sonnenschein, 1888. viii, 286 pp.

1118
[Blocquel, S.]. Le grand etteillia, or l'art de tirer les cartes . . . le tout recueilli, mis dans un nouvel ordre et corrigé, par Julia Orsini, sybille du Faubourg Saint-Germain.

Paris, chez les Marchands de Nouveautés, [n.d.]. 212, 78 (of illus.), illus., plate.

1119 a d

Bridges, F. Phrenology made practical and popularly explained. London, Liverpool, George Philip, 1860. xvi, 188 pp., illus.

1120 k

Bryan, B. The British vivisectors' directory: a black book for the United Kingdom; with a preface by Frances Power Cobbe. London, Society for the Protection of Animals from Vivisection; Swan Sonnenschein, 1890. vii, 102 pp.

1121

Clark, F. Le Gros. Manuals of elementary science; physiology. London, Society for Promoting Christian Knowledge, [etc.], 1873. 126 pp., front., illus.

1122 a d

Collyer, R. H. Automatic writing; The Slade prosecution; Vindication of the truth. London, H. Vickers, 1876. 23 pp. [*see* p.].

1123 i

— Exalted states of the nervous system in explanation of the mysteries of modern spiritualism, dreams . . . and nervous congestion; third edition. London, Henry Renshaw, 1873. iv, 144 pp. (Vienne, 3 May 1873).

1124 a e i

Cornish, K. H. Cholera treatment and cure . . . a letter to the Right Hon. Lord Carlingford. London, Baillière, Tindall, and Cox, 1884. 16 pp. [*see* p.].

1125 a d

Desbarrolles, A. Les mystères de la main révelés et expliqués; quatrième édition. Paris, E. Dentu; Librairie de la Société des Gens de Lettres, 1862. xxiv, 624 pp., illus.

1126 d i

[Dickson, S.]. Memorable events in the life of a London physician; in three parts. London, Virtue, 1863. [iv], 276 pp.

1127

Dowie, J. Appendix to the foot and its covering. London, Robert Hardwicke, 1872. 205-87 pp. Reprinted from *The foot and its covering*; 2nd edition. [*see* p.].

1128

— Remarks on the loss of muscular power arising from the ordinary foot-clothing now worn. London, Robert Hardwicke, 1863. 24 pp. [*see* p.].

1129 i

Durand, F. G. Traité dogmatique et pratique des fièvres intermittentes. Paris, F. Savy, 1862. xi, 464 pp.

1130 d i

Durand-Fardel, M. Lettres médicales sur Vichy; troisième édition. Paris, Germer Baillière, [etc.], [1866]. [iii], 246 pp., tables. (A. Hanoteau).

1131 i

Dutrieux, P. J. Considérations générales sur l'ophthalmie communément appelée ophthalmie d'Edypte; suivies d'une note sur les opérations pratiquées à l'Ecole khédiviale des Aveugles, au Caire; avec une préface en forme de lettre à S. E. Riaz-Pacha . . . traduit en Arabe par Hassan-Bey Mahmoud. Le Caire, Imprimerie de l'Etat-Major Général, 1878. xv, 150 pp., plates. At head of title: Publications de l'Etat-Major Général égyptien. (De passage à Aden, 9 Sept. 1878).

1132

Edgelow, G. Cancer and simple tumours dispersed by electricity. London, Henry Renshaw, [1883]. 16 pp.

1133 a d

Ford, H. A. Observations on the fevers of the West coast of Africa. New York, Edward O. Jenkins, printer, 1856. 48 pp.

1134 a d

Georgii, A. The movement – cure. London, H. Baillière, 1852. 24 pp. [*see* p.].

1135 a

Giaxa, V. de. Piccola enciclopedia di medicina ed igiene ad uso delle famiglie. Trieste, Julius Dase, 1883. 376 pp.

1136

Gibbons, R. A. The baths of Hammam R'Hira, Algeria, in the treatment of chronic rheumatism and gout. Algiers, Remordet, printer, 1888. 6 pp. Reprinted from the *Lancet,* 10 June 1888. [*see* p.]. Four copies.

1137

Gratiolet, P. Conférence sur la physionomie en général, et en particulier sur la théorie des mouvements d'expression. Paris, 1865. 20 pp. Reprinted from *Moniteur scientifique-Quesneville,* 1 Feb. 1865. [*see* 250].

1138

[Great Britain. Parliament]. Report of the General Board of Health on the epidemic cholera of 1848 & 1849; presented to both Houses of Parliament. London, W. Clowes, printer, 1850. viii, 156 pp., front. (map), illus. [*see* p.].

1139 a d

Jacolliot, L. Le spiritisme dans le monde, l'initiation et les sciences occultes dans l'Inde et chez tous les peuples de l'antiquité. Paris, A. Lacroix, 1879. [iii], 364 pp., illus.

1140

Jaeger, G. Selections from essays on health-culture and the sanitary woollen system ... translated from the German. London, Dr Jaeger's Sanitary Woollen System, 1884. viii, 197 pp.

1141 k

James, R. Pharmacopoeia universalis: or, a new universal English dispensatory . . . third edition, with very large and useful additions, and improvements. London, T. Osborne, W. Strahan, [etc.], 1764. viii, 668, [28] pp., illus.

1142

Kiernan, F. The anatomy and physiology of the liver. London, 1833. [60] pp., plates. Reprinted from *Philosophical transactions of the Royal Society,* Vol. 123, pp. 711-70. [*see* pq.].

1143

[Kingsley, C.]. Hints to stammerers, by a Minute Philosopher. London, Longman, 1864. [i], 33 pp. Reprinted from *Fraser's magazine for town and country,* July, 1859. [*see* 229].

1144 a d

Lavater, J. C. Physiognomy: or, the corresponding analogy between the conformation of the features and the ruling passions of the mind; translated from the original work; twentieth edition. London, Thomas Tegg, [etc.], 1844. xii, 336 pp., front., plates.

1145 l

Lewins, R. Life and mind: on the basis of modern medicine; edited by "Thalasso-plektos". London, Watts, 1880. 72 pp.

1146

London Mesmeric Infirmary. Report of the eighth annual meeting . . . London, Mitchell, printer, 1857. 40 pp. [*see* 224].

1147

McGeary, J. Illness and its cure by natural means as practised at the Metropolitan Psychopathic Establishment, 26 Upper Baker Street, N.W. [London], printed for private circulation, 1882. viii, 36 pp. [*see* p.].

1148 a d i

Maclaren, A. Training, in theory and practice; second and enlarged edition. London, Macmillan, 1874. xii, 252 pp., illus., plates. (Oxford, June 1875).

1149 a d

Maclaren, T. Systematic memory: or, how to make a bad memory good, and a good memory better; second edition. London, F. Pitman, 1866. 55 pp.

1150 d

M'William, J. O. Medical history of the expedition to the Niger during the years 1841-2, comprising an account of the fever which led to its abrupt termination. London, John Churchill, 1843. viii, 287 pp., front., illus., plates.

1151

Malan, H. V. Vade mecum of the homoeopathic practitioner; new edition. London, Leath & Ross, [1847]. viii, 220 pp.

1152 a d h

Martius, C. F. P. Systema de materia medica vegetal brasileira: contendo o catalogo e classificação de todas as plantas brasileiras conhecidas . . . extrahida e traduzida . . . pelo desembargador Henrique Velloso d'Oliveira. Rio de Janeiro, Eduardo & Henrique Laemmert, 1854. 284 pp.

1153

Matheson, D. On some of the diseases of women, their pathology and homoeopathic treatment: being four lectures delivered at the London Homoeopathic Hospital. London, Leath & Ross, 1876. vi, 76 pp.

1154

Mattei, C., Count. Electro-homoeopathy: the principles of new science discovered; third enlarged and revised English edition. Bologna, Printing Office Mareggiani, 1891.

[vi], 249 pp., front. (port.), plates.

1155 a d
Millingen, J. G. Curiosities of medical experience . . . second edition, revised and considerably augmented. London, Richard Bentley, 1839. xvi, 566 pp., tables.

1156
Moore, Y. J. The Anglo-Turkish bath: or, the modern application of the ancient Roman therma as a hygienic, prophylatic, and therapeutic agent. London, Simpkin, Marshall, 1861. ix, 58 pp., front. [*see* 229].

1157 a d
Morgan, S. The text book for domestic practice: being plain and concise directions for the administration of homoeopathic medicines in simple ailments; third edition. Bath, Edmund Capper, [etc.], 1861. 199 pp.

1158
Munk, W. Euthanasia: or, medical treatment in aid of an easy death. London, New York, Longmans, Green, 1887. vii, 105 pp.

1159 a d
Mure, B. and J. V. Martins. Pratica elementar da homeopathia; precedida de um discurso contendo a historia da homeopathia; terceira edição. Rio de Janeiro, Typographia e Livraria Franceza, 1847. lxxxiv, [20], 439 pp. At head of title: Publicação do Instituto homeopathico do Brasil.

1160 k
Nightingale, F. Notes on nursing: what it is, and what it is not. London, Harrison, [1860]. 79 pp. Heavily annotated by IB.

1161 a
Pairman, R. The great sulphur cure brought to the test; and workings of the new curative machine proposed for human lungs and windpipes; twelfth edition. Edinburgh, Edmonston & Douglas, 1868. 55 pp. [*see* 229].

1162 k
Parke, T. H. Guide to health in Africa, with notes on the country and its inhabitants; with preface by H. M. Stanley. London, Sampson Low, Marston, 1893. xi, 175 pp.

1163
[Peterson, A. T. T.]. Essays from the unseen, delivered through the mouth of W.L., a sensitive, and recorded by A.T.T.P. London, printed for the author, 1885. x, 528 pp., front. (port.), plates.

1164 l
Richardson, W. L. The duties and conduct of nurses in private nursing: an address delivered at the Boston Training School for Nurses, June 18, 1886 . . . London, Field & Tuer, [etc.], 1887. 59, x pp., table.

1165
St. Mark's Hospital. Laying of the corner stone of St. Mark's Hospital, Harper: Cape Palmas, Liberia . . . 24th April 1859. Goshen, N.Y., Charles Mead, 1860. 24 pp. [*see* 238].

1166 a d h
Sansom, A. E. The arrest and prevention of cholera: being a guide to the antiseptic treatment, with new observations on causation. London, John Churchill, 1866. [ii], 133 pp.

1167 l
Schurig, M. Muliebria historico-medica, hoc est partium genitalium muliebrium consideratio physico-medico-forensis, qua pudendi muliebris partes tam externae, quam internae . . . selectis et curiosis observationibus traduntur. Dresdae, Lipsiae, apud Christophori Hekelii B. Filium, 1729. [viii], 384, [34] pp.

1168 d l
Sinnett, A. P. The occult world. London, Trübner, 1881. [vii], 172 pp.
Another copy.

1169 l
[Sowter, T.]. Beulah Spa hydropathic establishment, Upper Norwood, S.E. [London, 1871]. 13 pp. [*see* 229].

1170 a d i
[Stainton-Moses, W.]. Psychography: a treatise on one of the objective forms of psychic or spiritual phenomena, by "M.A. (Oxon.)". London, W. H. Harrison, 1878. 152 pp., illus. (publisher, Nov. 1878).
a d
Another copy.

1171 h i

Tedeschi, V. L'alimentazione della prima infanzia manuale ad uso delle famiglie. Trieste, Tipografia G. Werk, 1889. [*see* p.].

1172 d i

Tuckey, C. L. Psycho-therapeutics: or, treatment by sleep and suggestion. London, Baillière, Tindall & Cox, 1889. xxi, 80 pp., illus. (April 1889).

1173 a d i

Turner, D. W., compiler. Rules of simple hygiene; seventh edition. London, Longmans, Green, 1874. [ii], 17 pp. (8 Oct. 1876). [*see* p.].

1174 a d h

Vago, A. L. Orthodox phrenology; second edition. London, Simpkin, Marshall, 1871. 133 pp., illus.

1175 a d

Walker, A. de N. The prophylactic power of copper in epidemic cholera. London, William Austin, printer, 1883. 24 pp. [*see* p.].

1176 d

[Walsh, J. H.]. The handbook of manly exercises; comprising boxing, walking, running, leaping, vaulting . . . by "Stonehenge"; "Forrest", &c. London, Routledge, Warne, & Routledge, 1864. 64 pp., illus.

1177

Wedgwood, J. J. Progress of dentistry and oral surgery; fifth edition. London, Henry Kimpton, 1886. vii, 96 pp., illus.

1178 d

Weiss, J. The hand book of hydropathy, for professional and domestic use . . . London, James Madden, 1884. xii, 438 pp.

1179 a

Wilson, J. Popular lectures on the preservation of health and cure of chronic disease. No. 1; third edition. London, Baillière, 1859. [ii], 104 pp., front. [*see* 229].

1180

Wylde, J. A pocket glossary of terms employed in anatomy, medical botany, pharmacy, surgery, domestic medicine . . . new edition. London, Leath & Ross; Leamington, Leath & Woolcott, [?1890]. 192 pp., table.

1181 d

Zoellner, J. C. F. Transcendental physics: an account of experimental investigations from the scientific treatises of . . . translated from the German, with a preface and appendices, by Charles Carleton Massey; third edition. London, W. H. Harrison, 1885. xlviii, 266 pp., front., illus.

1182–316 Religion, Philosophy

1182
Anonymous. Ke ri buku di utsha o Fernandian [form of catechism. Clarence, Fernando-Po, Dunfermline Press, printer, 1844]. 16 pp. [*see* 231].

1183
— London parishes, containing the situation, antiquity, and re-building, of the churches within the bills of mortality . . . [and] an account of the rise, corruption, and reformation, of the Church of England. London, B. Weed, printer for W. Jeffery, 1824. [iv], 156 pp.

1184
— Mienge ma Yehova: hymns in the Dualla language. Cameroons, Baptist Mission Press, printer, 1859. 64 pp. [*see* 234].

1185
'Abd al-Malik, son of Maulvī Muhammad Sādiq. ed. Majmu'ah-i Khutab dar sanah-i 1267 [A collection of Friday sermons for the year 1267 A.H.; Arabic text]. Bombay, Matba Muhammadi, 1267 A.H. [1850 A.D.].

1186 l
Ahmad ibn Husain (Abu al-Shiyā'), al-Isbahānī. Kitab Abu Shiyā' mukhtasaran fi 'ilm al-fiqh 'alā madhhab al-Imām al-Shāfi'i [A compendium of Islamic law according to the Shafi'ī school (known also as Ghāyat al-ikhtisār, and al-Taqrīb); Arabic text]. [n.p.], 1271 A.H. [1854 A.D.]. 70 pp. [*see* p.].

1187
— al-Taqrīb fi al-fiqh 'alā madhhab al-Imām al-a'zam Muhammad ibn Idrīs al-Shāfi'ī [A compendium of Islamic laws according to the Shāfi'ī school (also known as Ghāyat al-ikhtisār); Arabic text]. [n.p.], 1285 A.H. [1868 A.D.]. 56 pp. [*see* p.].

1188
Alishan, L. M. S. Théodore, la salahounien martyr arménien; traduit par J. Hékimian. Venise, Impr. arménienne de Saint-Lazare, 1872. 45 pp. [*see* 260].

1189
'Alī ibn Muhammad, al-Bazdawī. Fiqh al-akbar li'l-Shaikh 'Ali al-Pazdawi [The principles of the Muhammadan faith; edited by Lord Stanley of Alderley; Arabic text]. [London], 1279 A.H. [1862 A.D.]. 32 pp. 7 copies.

1190
Anger, R. ed. Synopsis evangeliorum Matthaei Marci Lucae cum locis qui supersunt parallelis litterarum et traditionum evangelicarum irenaeo antiquiorum . . . Lipsiae, sumtibus Gebhardti et Reislandi, 1852. [7], 1, 276, xlviii pp., tables. [*see* 1231, Vol. 3, part 2; spine: V].

1191 d i
Arnold, E. Pearls of the faith or Islam's rosary . . . with comments in verse. London, Trübner, 1883. xiv, 319 pp. (Jan. 1884).

1192 d
Arundell, J. F., Baron Arundell of Wardour. The nature myth theory untenable from the scriptural point of view. London, Burns & Oates, [1877]. v, 13 pp. [*see* p.].

1193
— The secret of Plato's Atlantis. London, Burns & Oates, 1885. [iv], 108 pp.

1194 a d
— Tradition principally with reference to mythology and the law of nations. London, Burns, Oates, 1872. xxix, 431 pp., tables.

1195
Athanasius, Saint, Patriarch of Alexandria. The orations . . . against the Arians. London, Sydney, Griffith, Farran, [etc.], [1888]. 299 pp.
1196
Austin, P. Our duty towards animals: a question considered in the light of Christian philosophy. London, Kegan Paul, Trench, 1885. [iii], 53 pp. [*see* p.].
1197 b d g h i j
Badger, G. P. Muhámmad [and] Muhámmadanism. London, John Murray, 1882, 951–98 pp. Reprinted from *Dictionary of Christian biography,* Vol. 3. Interleaved in a holograph notebook with 24 preliminary and 22 pages at the end. (George Percy Badger, 21 Feb. 1885; 17 July 1886; n.d.).
1198 a d h
Beke, C. T. Origines biblicae: or researches in primeval history. Vol. 1. London, Parbury, Allen, 1834. xvi, 336 pp., front. (map).
1199 l
[Bennett, A. R.]. Einsiedeln "In the dark wood", or our lady of the hermits: the story of an alpine sanctuary. Einsiedeln, Switzerland, Charles & Nicholas Benziger; London, Burns & Oates, 1883. 207 pp.
1200
[Bethel Missions]. A correspondence resulting from an appeal to the British public for money, to support an American Bethel Camp, in Havre, France; [3rd edition]. [London], Onwhyn, [1861]. 30 pp. [*see* 224].
1201
Bhagavadgîtâ. The Bhagavad-gítá: or, a discourse between Krishna and Arjuna on divine matters; translated . . . by J. Cockburn Thomson. Hertford, Stephen Austin, 1855. cxxxviii, 158 pp.
1202 l
Bible – English. The Holy Bible; translated from the Latin Vulgate; diligently compared with the Hebrew, Greek, and other editions, in various languages, with annotations by the Rev. Dr. Challoner . . . [and] The New Testament. New York, Montreal, D. & J. Sadlier, [1845]. 793, 228 pp., front.
1203 a
—. Old Testament – Albanian (in Greek characters). Psaltiri, kthyerë pas Evraishtesë vjetërë shqip ndë gjuhë toskërishte prej, Konstantinit Kristoforidhit, Elbasanasit. Konstantinopol, Ndë shtypa-shkronjë të A. H. Bojaxhianit, 1868. [viii], 122 pp.
1204
— — Arabic. The poem of poems: a metrical Arabic version of the Book of Job, &c. by R. Hassoun. London, Frederic Straker, 1869. viii, [160] pp.
1205
— — Arawakan. Akpa ñwed Moses eke ekerede Jenesis. Glasgow, Dunn ye Wright, 1862. [iii], 84 pp.
1206
— — Dualla. Kalati ya Beboteri: the Book of Genesis. Cameroons, Mission Press, 1861. [ii], 51 pp. [*see* 239].
1207
— — — [Mienge ma David: Psalms of David. Cameroons, Western Africa, Baptist Mission Press], 1857. 112 pp. [*see* 234].
1208
— — Hebrew – Commentary on Job. Tedeschi, Moses Isaac ben Samuel. [Mosheh Yitehak Ashkenazi]. Ho'il mosheh: be'ur hadash 'al sefer 'iyuv [Commentary on the Book of Job, written in Trieste; Hebrew text]. Padua, Francisco Sakito, 5635 A.M. [1874]. 104 pp. [*see* p.].
1209 i
— — Mpongwe. The Books of Genesis, part of Exodus, Proverbs, and Acts; translated into the Mpongwe language, at the Mission of the A.B.C.F.M., Gaboon, West Africa. New York, American Bible Society, 1859. 435 pp. (Wm. Walker, Gaboon, April 1862).
1210
— — Pentateuch. [Turkish text]. [240] pp.
1211 a
— New Testament – Albanian (in Greek characters). 'Ē Kainē Diathēke tou Kuriou kai Sōstēros 'Emōn Iesou Hristou diglōssos toutesti Bllēnikē [*sic*] kai Albanikē, meros 2. Dhiata ere e zotit sonë kë na shpëtoi Iesou Hrishtoit mbë digjuhë do me thënë Gjërkishte e dhe Shkipëtartze; pjeshi dhitë. En Athēnais, tupois H. Nikolaidou Filadelfeōs, 1858. 2 vols.

1212
— — Arabic. The Gospel of Matthew in Arabic, printed with all the vowels, according to the simplified method of the Revd. Jules Ferrete. London, Samuel Bagster, [1836]. viii, [118] pp.
1213
— — Benga. San iam, ya Matiu e lĕndĕkidi. New York, Mission House, 1858. 113 pp. [*see* 231].
1214 a
— — — Sango iam, ya Mark e lĕndĕkidi: the Gospel according to Mark, translated into the Benga language by Rev. Jas. L. Mackey. New York, American Bible Society, 1861. [ii], 71 pp. [*see* 231].
1215
— — Dualla. Kalati e ta e loma na miemba: na kalati ya bebiisedi: Epistles to the churches, and the Book of Revelation. [Cameroons], Western Africa, Mission Press, 1861. [i], 193 pp. [*see* 234].
1216
— — — Kalati ya loba, bwambu ba Dualla: scriptures in the Dualla or Cameroons language. Cameroons, Western Africa, Mission Press, printers, 1857. [iii], 288 pp. [*see* 234].
1217
— — Efik. Öbufa Testament äbon ye andinyaña nyïn Jisus Krist, ke ifä Efïk. Edinburgh, Murray ye Gibb, [etc.], [National Bible Society of Scotland, 1862]. [iii], 403 pp. [*see* 1205].
1218
— — French. La vie de N.S. Jésus-Christ écrite par les quatre Evangélistes . . . rédigée et présentée . . . par M. l'Abbé Brispot. Paris, Plon, 1853. 2 vols., fronts., plates. [*see* q.].
1219
— — Grebo. The Acts of the Apostles, translated into the Grebo tongue, by the Rev. John Payne. New York, American Bible Society, 1851. 98 pp.
1220
— — — The Gospel according to St. John, translated into the Grebo tongue, by the Right Rev. John Payne. New York, American Bible Society, 1852. [ii], 79 pp.
1221 d
— — Greek. 'E Kainē Diathēkē tou Kuriou kai Sōtēros 'ēmoñ Iēsou Hristou, metafrastheisa ek tou 'Ellēnikou. En Kantabrigiaj, Etupōthē di' epimeleias . . . Tupothetou tēs Akadēmias: dapanēj tēs 'Ierografikēs 'Etaireias pros diadosin tou Theiou Logou eis te tēn Bretannian kai ta alla ethnē, 1869. [iii], 479 pp.

1222 a
— — Icelandic. Hiđ Nýa Testamenti drottins vors Jesú Krists; ásamt međ Davíđs Sálmum; endurskođuđ útgáfa. Oxford, prentađ í Prentsmiđju Háskólans í Oxford, á Kostnađ hins Brezka og Erlenda Biflíufèlags, 1866. [iii], 628 pp.
1223 a
— — Irish. Tiomna Nuadh, Ar Dtighearna agus Ar Slanuigheora, Josa Criosd; ar na tarruing go firinneach as Greigis go Gaoidheilg re Huilliam O Domhnuill. A Lunnduin, ar na cur a gclo re Robert Ebheringtam, an bliadain d ois an Tigerna, 1681. [i], 364 pp.
1224
— — Roumanian. Noul Testament al domnuluĭ şi mântuitoruluĭ Nostru Iisus Christos; publicat de Societatea biblică pentru Britania şi străinătate. Jassy, Tipografia H. Goldner, 1871. [ii], 354 pp.
1225
— — Russian. Novyy Zavet [Gospoda] nashego Iisusa Khrista. N'yu Iork, 1867. [i], 607 pp.
1226 a d h
— — Slovenian. Novi Zakon Gospoda in zveličarja našega Jezusa Kristusa I. del. Čveteri Evangelji in dejanja Sv. Aposteljnov; poleg grškega izvirnika. Na Dunaji, Založil A. Reichard in Druž, 1873. 321 pp.
1227 d
— — Turkish. al-Injil Sharif, Incili Serif. [Turkish translation of the New Testament]. London, British & Foreign Bible Society, 1854. 345 pp.
1228 i
— — Zingaresco. Il cantico dei cantici di Salomone per la prima volta tradotto dal testo italiano in fronte nell'idioma Zingaresco (Indo-orientale): studio di James

Pincherle. Trieste, Stablimento Tipografico e Calcografico del "Tergesteo" di Giovanni Balestra, 1875. 14 pp. [*see* 220].

1229 l

— — Apocrypha. The Apocryphal New Testament: being all the Gospels, Epistles, and other pieces now extant, attributed in the first four centuries to Jesus Christ, His Apostles . . . London, William Hone, 1820. 271 pp.
Another copy.

1230 b

— Polyglott. The Bible of every land: a history of the sacred scriptures in every language and dialect into which translations have been made . . . new edition. London, Samuel Bagster, [1860]. [xxxii], 480 pp., plates, maps.

1231 a b d

— — Polyglotten-Bibel zum praktischen Handgebrauch: die heilige Schrift Alten und Neuen Testaments in übersichtlicher Nebeneinanderstellung des Urtextes, der Septuaginta, Vulgata und Luther-Uebersetzung . . . bearbeitet von R. Stier und K. G. W. Theile. Beilefeld, Velhagen und Klasing, 1847–55. 4 vols. (in 6).

1232

— — St. John iii, 16 in most of the languages and dialects in which the British & Foreign Bible Society has printed or circulated the Holy Scriptures; enlarged edition. London, British & Foreign Bible Society, 1878. [ii], 48 pp.

1233

— — St. Johannis iii, 16. in den meisten der Sprachen und Dialecte in welchen die britische und ausländische Bibel-Gesellschaft die heilige Schrift gedruckt und verbreitet hat; vermehrte Ausgabe. London, Britische und ausländische Bibel-Gesellschaft, 1878. [ii], 48 pp.

1234 a d

Bigandet, P. The life, or legend of Gaudama, the Budha of the Burmese, with annotations; the ways to Neibban, and notice on the phongyies, or Burmese monks. Rangoon, American Mission Press, C. Bennett, 1866. xi, 538, v pp.

1235

Blavatsky, H. P. Isis unveiled: a master-key to the mysteries of ancient and modern science and theology; second edition. New York. J. W. Bouton; London, Bernard Quaritch, 1877. 2 vols., illus., plate.

1236 d l

Bocharto, S. Hierozoicon sive bipertitum opus de animalibus sacrae scripturae. Londini, excudebat Tho. Roycroft, [etc.], 1663. [48] pp., 1094 columns; [124] pp.; 888 columns; [118] pp., front., illus. At head of annotated columns: De Phoenice. [*see* q.].

1237

Book of Mormon, The. The book of Mormon: an account written by the hand of Mormon, upon plates taken from the plates of Nephi; translated by Joseph Smith, jun.; fourth European edition, stereotyped. Liverpool, S. W. Richards for Orson Pratt, 1854. [iii], 563 pp.

1238 a

British National Association of Spiritualists, The. [Constitution and rules as amended by the Council]. London, [1876]. 15 pp. [*see* p.].

1239 a

Butler, J. The analogy of religion, natural and revealed, to the constitution and course of nature . . . with a preface by Samuel Halifax; new edition. London, Henry G. Bohn, 1856. vi, 546 pp. (Bohn's standard library).

1240

Caddell, C. M. A history of the missions in Paraguay. London, Burns & Lampert, 1862. iv, 102 pp. Title-page wanting.

1241 a

Caithness, M. S., Countess of; Duchesse de Pomár. The mystery of the ages contained in the secret doctrine of all religions. London, C. L. H. Wallace, 1887. xxxii pp. Preface and contents only. [*see* p.].

1242

Calza, G. Saggio sulla religione de' Maomettani. In Venezia, Stamperia Presso Antonio Fortunato Stella, 1794. xxiv, 173 pp.

1243 e

Clarke, H. Serpent and Siva worship and mythology in central America, Africa, and Asia. London, 1876. 14 pp. Reprinted from the *Journal of the Anthropological Institute,* Vol. 6. [*see* p.].

1244 a d h
— Ten great religions: an essay in comparative theology. Boston, James R. Osgood, 1876. x, 528 pp., front.

1245
Clemens, W. Nuwe j ipakua ja ejanganangobo ya anyambe; scripture questions in the Benga language. New York, Mission House, 1861. 117 pp. [*see* 231].

1246 l
Dupanloup, F. Remarks on the encyclical of the 8th of December, A.D. 1864; translated . . . by William J. M. Hutchison. London, Geo. Cheek, 1865. [iii], 60 pp. [*see* p.].

1247 a d
Dupin, A. M. J. J. Jésus devant Caïphe et Pilate: ou procès de Jésus-Christ suivi d'un choix de textes contenant les principaux fondements de la religion chrétienne. Paris, chez F.-H. Barba, 1864. xiii, 303 pp., illus.

1248
— Juicio critico del proceso de Jesucristo. Santa Cruz de Tenerife, Salvador Vidal, 1864. 47 pp. [*see* 226].

1249 d
Ferreira Machado, S. Triumpho eucharistico: exemplar da Christandade lusitana em publica exaltação de fé na solemne trasladação do divinissimo sacramento da igreja da Senhora do Rosario . . . Lisboa occidental, na Officina da Musica, [etc.], 1734. 44, 6 pp. [*see* 1770].

1250
Figuier, L. The day after death: or, our future life, according to science; translated from the French; a new . . . edition. London, Richard Bentley, 1884. viii, 308 pp., illus., plates.

1251 e i
[Garbett, E. L.]. The ascertainable in religion; seven miracles identifying the church. London, George J. Stevenson, 1871. 32 pp. [*see* 1933].

1252 d i l
Gomez Quevedo Villegas, F. de. The visions of Dom Francisco de Quevedo Villegas . . . made English by Sir Roger L'Estrange; the eleventh edition, corrected. London, printed by W. B. for Richard Sare, 1715. 140 pp. (C. Carter Blake).

1253
[Gray, E. McQ.]. The Oberammergau passion play, 1880: arrangements for visitors. London, Aberammergau, E. McQueen Gray, 1880. 8 pp. [*see* 235].

1254 j
— To Oberammergau and back: a practical guide for visitors to the passion play, 1880. London, Obberammergau, published by the author, 1880. 40 pp. (L. Clifton Lyne, [London], 20 April 1880). [*see* 235].

1255 a d
Hardy, R. S. A manual of Budhism, in its modern development; translated from Singhalese mss. London, Edinburgh, Williams & Norgate, 1860. xvi, 534 pp.

1256
Hermes, Trismegistus. The virgin of the world of Hermes Mercurius Trismegistus; now first rendered into English with essay, introductions and notes by Dr. Anna Kingsford and Edward Maitland. London, George Redway, 1885. xxxii, 154 pp., front., illus. At head of title: The Hermetic works. imperfect.

1257
Hitōpadēsa. Akhlāq-i Hindi [Indian ethics; an Urdu translation by Mīr Bahādūr 'Alī, of the Persian translation, called "Mufarrih al-qulūb", by Tāj al-Dīn ibn Mu'īn al-Dīn Malikī, of a Hindi version of the Hitōpadēsa, a Sanskrit book of fables; translated under the superintendance of James Gilchrist; Urdu text. Bombay, 1835]. 342 pp.

1258 e l
[Holland, H. S.]. Impressions of the Ammergau passion-play, by an Oxonian. London, J. T. Hayes, 1870. 32 pp. [*see* 235].

1259 l
Hone, W. Ancient mysteries described, especially the English miracle plays, founded on Apocryphal New Testament story, extant among the unpublished manuscripts in the British Museum. London, William Hone, 1823. 299 pp., front., illus., plates.

1260 a d h
Hughes, T. P. A dictionary of Islam: being a cyclopaedia of the doctrines, rites, ceremonies and customs, together with the technical and theological terms, of the Muhammadan religion. London, W. H. Allen, 1885. viii, 750 pp., illus., map.

1261 a d

— Notes on Muhammadanism. London, Wm. H. Allen, 1875. xv, 208 pp., tables.

1262

Isacchi, A. and T. Isacchi. Meditazioni per ciascun giorno del mese e guida per ascoltare la Santa messa; contadine illetterate della Provincia di Como. Firenze, Tip. Cattolica diretta da G. Papini, 1869. xxiv, 300 pp.

1263 d l

Jackson, J. P. The Ober-ammergau passion play. London, sold by W. H. Smith, [etc.], 1880. vii, 86 pp., front., plates, illus. [*see* 235].

1264 k

Koran. Arabic. [Qur'ān; Arabic text. Constantinople?], 1290 A.H. [1873 A.D.]. 30 parts.

1265 l

— — Coranus arabice: recensionis flügelianae, textum recognitum iterum exprimi curavit Gustavus Mauritius Redslob; editio stereotypa. Lipsiae, typis et sumtu Caroli Tauchnitii, 1837. viii, [540] pp.

1266 a d h i

— English. The Koran, commonly called the Alcoran of Mohammed; translated from the original Arabic with explanatory notes . . . by George Sale; a new edition. London, printed for Charles Daly, 1838. xvi, 134, 472 pp., front., plates, map. (Wm Morley, Oct. 1861).

1267 a d

— — El-Kor'ân: or, the Korân; translated from the Arabic, the Suras arranged in chronological order; with notes and index by J. M. Rodwell; second revised and amended edition. London, Bernard Quaritch, 1876. xxviii, 562 pp.

1268

— French. L'Alkoran! le livre par excellence; traduction textuelle de l'Arabe faite par Fatma-Zaïda Djarié-Odalyk-Doul den Béniamïn-Aly Effendi-Agha. Lisbonne, Imprimerie de la Société typographique franco-portugaise, 1861. 483, viii pp.

1269

Kremer, A. von. Culturgeschichtliche Streifzüge auf dem Gebiete des Islams. Leipzig, F. A. Brockhaus, 1873. 77 pp. [*see* 257].

1270 a d

— Mollâ-shâh et le spiritualisme oriental. Paris, Imprimerie impériale, 1869. [iv], 55 pp. [*see* 258].

1271 d

— Notice sur Sha'râny. [Paris, Imprimerie impériale], 1868. 19 pp. Reprinted from the *Journal asiatique,* Series 6, Vol. 11. [*see* 257].

1272 l

La Croix de Ravignan, G. F. X. On the Jesuits, their institute, doctrines &c. &c., translated from the French . . . by W. C. Atchison. London, John Ollivier, 1844. xvi, 100 pp. [*see* 225].

1273

Lewes, G. H. Comte's philosophy of the sciences: being an exposition of the principles of the Cours de philosophie positive of Auguste Comte. London, George Bell, 1875. viii, 351 pp. (Bohn's philosophical library).

1274 a

Lewins, R. ed. What is religion? a vindication of freethought by C.N.; annotated by Robert Lewins. London, W. Stewart, 1883. 63 pp. [*see* p.].

1275 k

Locke, J. The conduct of the understanding; a new edition, divided under heads. London, printed by J. Cundee for M. Jones, 1802. [iv], 162 pp.

1276 a d h

McCaul, A. The old paths: or, a comparison of the principles and doctrines of modern Judaism, with the religion of Moses and the Prophets. London, London Society's House, 1854. xii, 476 pp.

1277 h l

MacColl, M. The Ober-ammergau passion play; reprinted, by permission, from the "Times"; with some introductory remarks on the origin and development of miracle plays; new and revised edition. London, [etc.], Rivingtons, 1880. viii, 104 pp.

1278 a d h

M'Taggart, W. B. An examination and popular exposition of the hylo-idealistic philosophy. London, W. Stewart, [1884]. vii, 84 pp. Pamphlets inserted: Evolution, life, the soul, by Inquirer, London, 1886. 15 pp. – Humanism versus theism . . . in a series of letters by Robert Lewins, London, 1887. 32 pp.

1279 l
Makhat, M. and P. Nemer. Elementarbuch [Arabic text, with some German]. Wien, Gedruckt in der K. K. Hof- und Staatsdruckerei zu Wien, 1862. 134 pp., plates.

1280 a d l
Massey, G. Concerning spiritualism. London, James Burns, [1872]. viii, 120 pp.

1281 d
Mill, J. S. A system of logic, ratiocinative and inductive, being a connected view of the principles of evidence, and the methods of scientific investigation; fourth edition. London, John W. Parker, 1856. 2 vols.

1282 l
Molloy, G. The passion play at Ober-ammergau, in the summer of 1871; second edition. London, Burns, Oates, [etc.], 1872. 120 pp., front., plates.

1283 a d h
Moor, E. The Hindu pantheon; a new edition . . . condensed and annotated, by W. O. Simpson. Madras, J. Higginbotham, [etc.], 1864. xvi, 401 pp., front., plates.

1284 d l
[Moore, T.]. Travels of an Irish gentleman in search of a religion, with notes and illustrations by the Editor of "Captain Rock's Memoirs"; second edition. London, Longman, Rees, [etc.], 1833. 2 vols.

1285
Muñoz Maldonado, J., Count de Fabraquer. Las catacumbas ó los martires: historia de los tres primeros siglos del Cristianismo. Madrid, Establecimiento Tipografico de D. F. de P. Mellado, 1849. xv, 360 pp.

1286
New York Sun. A celestial Utopia: extracted from the "New York Sun" of April 30th, 1869. [Frome, London, Butler & Tanner, printers], 1869. 16 pp. [*see* 228].

1287
Nierses Clajensis, St. Preces sancti Nersetis Clajensis Armeniorum Patriarchae Triginta; tribus linguis editae. Venetiis, in Insula S. Lazari, 1862. [ix], 552 pp., front.

1288
Parkhurst, J. A Greek and English lexicon to the New Testaments . . . to this work is prefixed, a plain and easy Greek grammar . . . a new edition, carefully revised . . . by J. R. Major. London, Longman, [etc.], 1851. xxviii, 720 pp., plate.

1289 i
Peebles, J. M. Witch-poison and the antidote, or Rev. Dr. Baldwin's sermon on witchcraft, spiritism, hell and the devil re-reviewed. Troy, N.Y., Troy Children's Progressive Lyceum, 1872. 94 pp. [*see* p.].

1290 k
Phelps, E. S. Beyond the gates. London, Chatto & Windus, 1883. [iv], 187 pp.

1291 a d h
Philo, Judaeus. The works of Philo Judaeus, the contemporary of Josephus; translated from the Greek, by C. D. Yonge. London, Henry G. Bohn, 1854–5. 4 vols. (Bohn's ecclesiastical library).

1292
Pincherle, J. Scenic narratives from the Bible. Trieste, Weis, printer, 1870. 31 pp. [*see* p.].

1293
Plato. Platōnos Apologia Sōkratous [with supplementary material; Greek text]. En Londinōj, Clayton, printer, 1873. [iv], 64 pp. [*see* p.].

1294 d f
Proctor, R. A. The religion of the Great Pyramid. [London], 1877. 331–43 pp. From an unidentified journal. [*see* 60].

1295
Redway, G. Bookseller. The literature of occultism and archaeology: being a catalogue of books on sale . . . December 1885. [London], 1885. 48 pp. [*see* p.].

1296 a h
Renan, E. Le Judaïsme comme race et comme religion conférence faite au cercle Saint-Simon, le 27 janvier 1883; [deuxième édition], 1883. [iii], 29 pp. [*see* p.].

1297 a d
Schenkel, D. A sketch of the character of Jesus: a Biblical essay; translated from the third German edition. London, Longmans, Green, 1869. xvi, 395 pp.

1298
Schoeberl, F. Das Oberammergauer Passions-Spiel mit den Passionsbildern von

A. Dürer. Eichstätt, München, Krüll, [1880]. [iii], 86 pp., illus., map. [*see* p.].

1299 j

Singer, J. Sollen die Juden Christen werden? ein offenes Wort an Freund und Feind; zweite . . . Auflage. Wien, Oskar Frank, 1884. xv, 143 pp. (J. Singer, Wien, 17 Juni 1884). [*see* p.].

1300 a d l

Smith, W. ed. A concise dictionary of the Bible, for the use of families and students. London, John Murray, 1865. [iv], 1039, illus., maps.

1301 a d h

Smith, Sir W. R. The Old Testament in the Jewish church; twelve lectures on Biblical criticism. Edinburgh, Adam & Charles Black, 1881. xii, 446 pp.

1302 l

Society for Promoting Christian Knowledge. Committee of General Literature and Education. The scripture atlas for the use of schools. London, [*c.* 1860]. 8 pp., plan, maps. [*see* p.].

1303 h

Society of Jesus. Collections towards illustrating the biography of the Scotch, English, and Irish members, S. J. Exeter, W. C. Featherstone, printer, 1838. xiv, 263 pp.

1304 k

[—]. A short account of the declaration given by the Chinese Emperour Kam Hi, in the year 1700. London, 1703. [xxviii], 71 pp.

1305 a d

Spinoza, B. de. Tractatus theologico-politicus: a critical inquiry into the history, purpose, and authenticity of the Hebrew scriptures . . . from the Latin; with an introduction and notes by the editor. London, Trübner, 1862. viii, 359 pp.

1306

Spiritual gifts [chapter on]. pp. 65–80. title-page wanting. [*see* 243].

1307

Stead, W. T. The passion play as it is played to-day, at Ober Ammergau in 1890. London, Office of the "Review of Reviews", [etc., 1890]. 130 pp., front., illus. [*see* p.].

1308 l

Stolz, A. Das Vaterunser und der unendliche Gruss; fünfzehnte Auflage. Freiburg im Breisgau, Herder'sche Verlagshandlung, 1885. 4 pts. (in 1), illus. (Gesammelte Werke, 4).

1309 a d

Swedenborg, E. Heaven and hell; also, the intermediate state: or world of spirits, a relation of things heard and seen. London, Swedenborg Society, 1863. [viii], 370 pp.

1310 a d h l

Talmud. Traité des Berakhoth du Talmud de Jérusalem et du Talmud de Babylone; traduit pour la première fois en français par Moïse Schwab. Paris, Imprimerie Nationale, 1871. vii, 560 pp.

1311

Tedeschi, M. ed. I primi profeti commentati ad uso delle Scuole israelitiche da Moisè Tedeschi, istruttore nel Talmud-Torà di Trieste; [Hebrew text]. Gorizia, Tip. Seitz, 1870. 302 pp.

1312

— Libro d'istruzione religiosa ad uso delle classi inferiori nelle Scuole israelitiche; seconda edizione. Venezia, Trieste, Stab. Tip. di C. Coen, 1872. [iii], 95 pp. [*see* p.].

1313 k

Thompson, Mrs B. Kitāb al-haiwat al-fādī al-majid; [Life of the Blessed Redeemer; translated by Salīm Bustani; Arabic text]. Beirut, Matba'at al-Ma'ārif, 1869. 244 pp., front.

1314 k

Thompson, W. H. Early chapters – in – Hull Methodism, 1746–1800. London, Charles H. Kelly, [etc.]. 1895. vii, 77 pp., plates.

1315

Vay, C., and others, eds. Geist, Kraft, Stoff; herausgegeben von Catharina, Adelma und Odön Vay. Wien, Adolf Holzhausen, 1870. viii, 168 pp., illus. [*see* p.].

1316 d l

Wyl, W. Maitage in Oberammergau: eine artistische Pilgerfahrt . . . mit dem zum ersten Male veröffentlichten Texte des Passionsdrama's . . . Zürich, Cäsar Schmidt, 1880. 2 vols. (in 1), front., plates.

1317-407 Sciences, Pure, Natural, Applied

1317 d
Anonymous. The Niger trade. [Manchester?, Shirrefs & Russell Mona Steam Press, printers, 186–?]. 71 pp., front. (map). [*see* 227].

1318 l
— al-Ruznāmat al-Sūrīyah, ai al-munākh al-sūrī. [An almanac of Syria; Arabic text. Beirut, 1868]. [24] pp.

1319
— Il solitario sicano di terme-selinuntina astronomo del monte cronio ovvero caledario profetico per l'anno 1886, anno decimonono. Sciacca, Redazione del Solitario Sicano, [1885]. 64 pp., illus.

1320 l
Accademia di Commercio e Nautica in Trieste. Osservazioni meteorologiche dell'I.R. Accademia di Commercio e Nautica in Trieste; mese di aprile 1872 [-novembre 1872]. [Trieste], 1872. 8 sheets [*see* p.].

1321 a l
Aḥmad ibn Moḥammad ibn Kathīr, al-Farghānī. Muhamedis Alfragani Arabis chronologica et astronomica elementa, e palatinae bibliothecae veteribus libris versa, expleta, & scholis expolita; additus est commentarius . . . autore M. Iacobo Christmanno. Francofurdi, apud Andreae Wecheli heredes, 1590. [xiv], 566 pp., tables.

1322 a d
Anderson, A. A. Terra: on a hitherto unsuspected second axial rotation of our earth . . . London, Reeves & Turner, 1887. xv, 175 pp.

1323 j k
— — second edition, with additional geological facts, and replies to criticisms . . . London, Reeves & Turner, 1890. xv, 253 pp. (Cranleigh, Guildford, 20 Nov. 1893 to IB).

1324 d
Ansted, D. T. An elementary course of geology, mineralogy, and physical geography. London, John van Voorst, 1850. xxvii, 584 pp., illus.

1325
Ashworth, H. Cotton: its cultivation, manufacture, and uses; a paper read before the Society of Arts, London, 10 March 1858. Manchester, James Collins, printer, [1858]. 64 pp. [*see* 227].

1326 l
Baumgartner, A., compiler. Trigonometrisch bestimmte Höhen von Osterreich, Steiermark . . . mit Einschluss des Görzer und Triester Kreises. Wien, Carl Gerold, 1832. [i], 101 pp. Reprinted from *Zeitschrift für Physik und Mathematik,* Vol. 10. [*see* p.].

1327 d
Beudant, F. S. Cours élémentaire d'histoire naturelle . . . minéralogie . . . cinquième édition. Paris, Langlois et Leclercq, Victor Masson, 1851. 2 vols. (in 1), illus. Sub-title of Vol. 2: Geologie.

1328 d
Bianconi, G. G. Esperienze intorno alla flessibilità del ghiaccio: memoria. Bologna, Tipi Gamberini e Parmeggiani, 1871. 14 pp., plates. Reprinted from the *Memorie dell'Accademia delle Scienze dell'Istituto di Bologna,* Series 3, Vol. 1. [*see* p.].

1329 a d
Bill, J. G. Grundriss der Botanik für Schulen; vierte, umgearbeitete Auflage. Wien,

Druck und Verlag von Carl Gerold's Sohn, 1866. viii, 264 pp., illus. [*see* 1390].

1330

Bingley, W. Useful knowledge: or a familiar account of the various productions of nature, mineral, vegetable, and animal; second edition, with considerable alterations and additions. London, Baldwin, Cradock, and Joy, [etc.], 1818. 2 vols., fronts., plates, tables.

1331 e (p. 25)

Bland & Co. Reduced price list and illustrated catalogue of apparatus & chemical preparations used in the art of photography. London, Bland, 1859. 75 pp. [*see* 223].

1332 d l

Bolle, G. Le malattie del baco da seta. [Gorizia], Tip.-Seitz ed., [1874]. 22 pp. Reprinted from *L'Isonzo*. [*see* p.].

1333 a d

Buckland, F. T. Fish hatching. London, Tinsley, 1863. xv, 268 pp., front., illus.

1334 a

Bussolin, G. Della Imperiale Privilegiata Compagnia orientale nel secolo scorso e del Lloyd Austro-Ungarico nel secolo presente . . . Trieste, Stab. Tipogr. di Lod. Herrmanstorfer, 1882. vii, 246 pp.

1335 i

Buzzi, L. Riflessioni sulle ferrovie italiane per l'ingegnere. Trieste, Tipografia Morterra, 1874. 87 pp. [*see* 230].

1336 d

Caravia, A. T. Cultura do algodão . . . publicado no diario do Rio Grande; offercido aos agricultores da Provincia do Rio Grande do Sul, por R. C. Dillon. Manchester, 'Guardian' Steam-Printing Offices, 1864. 12 pp. [*see* 227].

1337 i

Casella, L. An illustrated and descriptive catalogue of surveying, philosophical, mathematical, optical, photographic, and standard meteorological instruments, manufactured by L. Casella, scientific instrument maker to the Admiralty. London, [1871]. vii, 260 pp., illus. (23 Aug. 1871).

1338

Chemin de Fer de Constantinople jusqu'aux limites de la servie, avec embranchement a Salonique. Londres, W. Clowes, imprimerie, 1860. 39 pp. At head of title: Sublime Porte. [*see* 245].

1339

Coleman, M. Stenography: or a brief and simple system of short-hand. London, W. & H. S. Warr, 1857. 24 pp., plates. [*see* 223].

1340

Cotton Supply Association. Cotton culture in new or partially developed sources of supply; report of proceedings at a conference held on Wednesday, August the 13th 1862. Manchester, John J. Sale, printer, 1862. 56 pp. [*see* 227].

1341 l

— The cultivation of Orleans staple cotton, from the improved Mexican cotton seed as practised in the Mississippi cotton-growing region; third edition. [Manchester, 'Guardian' Steam-Printing Offices, 1864]. 34 pp., illus. [*see* 227].

1342

— El cultivo del algodon llamado Nueva Orleans, producido de la semilla mejicana mejorada del modo practicado en le region algodonera del Mississippi. Manchester, Juan J. Sale, impresor, 1861. 39 pp., front., illus. [*see* 227].

1343 a d

Dana, J. D. and G. J. Brush. A system of mineralogy: descriptive mineralogy, comprising the most recent discoveries; fifth edition, re-written and enlarged. London, Trübner; New York, John Wiley, 1871. xlviii, 827 pp., illus.

1344 a d l

Davies, C. Elementary algebra: embracing the first principles of the science. New York, A. S. Barnes, 1859. viii, 303 pp., table.

1345

Dorn, A. Die Eisenbahnpolitik der Zukunft. Triest, Selbstverlag des Verfassers; Buchdruckerei des Oesterr. Lloyd, printer, 1872. 18 pp. Reprinted from *Triester Zeitung*, 7–10 Aug. 1872. [*see* p.].

1346

[Dossie, R.]. The handmaid to the arts . . . second edition, with considerable additions and improvements. London, J. Nourse, 1764. 2 vols.

1347 i
Dowie, J. The foot and its covering; comprising a full translation of Dr. Camper's work on "The best form of shoe"; second edition. London, Robert Hardwicke, 1871. xxviii, 278 pp., illus., plates.

1348
Fenwick de Porquet, L. Foreign and English and English and foreign ready reckoner of monies, weights, and measures, for nearly all parts of Europe . . . seventh edition. London, Simpkin, Marshall, 1868. [vi], 54 pp.

1349 l
Forskål, P. Flora aegyptiaco-arabica, sive descriptiones plantarum, quas per Aegyptum inferiorem et Arabiam felicem . . . post mortem auctoris edidit Carsten Niebuhr . . . Hauniae, ex officina Mölleri, 1775. [2], 32, cxxvi, 220 pp., front. (map).

1350 d
Fothergill, J. M. Food for the invalid: the convalescent, the dyspeptic, and the gouty; second edition, revised and enlarged. London, Macmillan, 1884. [viii], 167 pp., plate.

1351 i
[Freeland, H. W.]. On the advantages of good roads to a country; [English and Arabic text]. [London, W. M. Watts, printer, 1862]. [vi], 10, [12] pp. [*see* p.].

1352 a d
Geologia elementar ou manual de geologia. [Rio de Janeiro. 18–?]. [ii], 136 pp. wanting title-page. [*see* 242].

1353 k
Gerard, J. The herball: or generall historie of plantes; very much enlarged and amended by Thomas Johnson . . . London, Adam Islip Joice Norton & Richard Whitakers, printers, 1636. [xxxiii], 1630 pp., illus., plate. [*see* q.].

1354 a d j
Goslin, S. B. The relative advantages of wind, water, and steam as motive powers, compared with each other . . . second edition. London, John Warner, & M'Corquodale, 1881. 50 pp., illus. (James Irvine, Liverpool, 22 Oct. 1881). [*see* p.].

1355
Gomes, F. L. De la question du coton en Angleterre et dans les possessions portugaises de l'Afrique occidentale. Lisbonne, Imprimerie de la Société typographique franco-portugaise, 1861. 34 pp. [*see* 227].

1356
Gordon, J. Lunar & time tables, adapted to new, short, and accurate methods for finding the longitude by chronometers and lunar distances . . . third edition. London, James Imray, & J. D. Potter, 1853. xvi, 203 pp., illus., maps.

1357 a d
Great Britain. Geological Survey of the United Kingdom. Museum of Practical Geology. A descriptive guide to the Museum . . . with notices of the Geological Survey of the United Kingdom, the Royal School of Mines, and the Mining Record Office, by Robert Hunt and F. W. Rudler; third edition. London, George E. Eyre & William Spottiswoode, printer for Her Majesty's Stationery Office, 1867. viii, 167 pp.

1358 d l
Grove, Sir W. R. The correlation of physical forces . . . fifth edition; followed by a discourse on continuity. London, Longmans, Green, 1867. xvii, 363 pp.

1359 j
Guenther, A. On new species of snakes in the collection of the British Museum [and] On new species of batrachians from Australia. London, 1863. 10 pp., plates. From *Annals and magazine of natural history,* Series 3, Vol. 2. (autograph list of shells presented to the British Museum by Lady Burton, signed W. Baird, 27 Dec. 1862; 29 Dec. 1862; J. E. Gray, 2 Jan. 1862). [*see* 225].

1360 a d
Herschel, Sir J. F. W. ed. A manual of scientific enquiry, prepared for the use of officers in Her Majesty's Navy, and travellers in general; second edition. London, John Murray, 1851. xi, 504 pp., illus., maps.

1361 a d i
Humboldt, F. H. A. von. Correspondence scientifique et littéraire recueillie, publiée et précédée d'une notice et d'une introduction par M. de la Roquette . . . suivie de la biographie des correspondants de Humboldt, de notes et d'une table . . . Paris, E. Ducrocq, 1865. xliv, 466 pp., illus., plates.

1362
Huxley, T. H. Physiography: an introduction to the study of nature. London, Macmillan, 1878. xx, 384 pp., front. (map), illus., plate, maps.

1363

Inman, J. Navigation and nautical astronomy, for the use of British seamen; thirteenth edition, revised. London, Rivingtons, 1862. iv, 280 pp., illus.

1364

Joncourt, M. de. Wholesome cookery. London, Kegan Paul, Trench, 1882. xiii, 188 pp.

1365

Jukes, J. B. and others. Lectures on gold for the instructions of emigrants about to proceed to Australia . . . delivered at the Museum of Practical Geology. London, David Bogue, 1852. vii, 215 pp., illus. [*see* 1327].

1366 l

Laplace, P. S. de. Elementary illustrations of the celestial mechanics. London, John Murray, 1832. [v], 344 pp., illus.

1367 d h

Liais, E. L'espace céleste et la nature tropicale: description physique de l'univers d'après des observations personnelles faites dans les deux hémisphères; préface de M. Babinet . . . Paris, Garnier Frères, [1866]. xiii, 606 pp., front. (port.), illus., plates, map.

1368 a

Lilly, W. An introduction to astrology; with numerous emendations, adapted to the improved state of the science in the present day; A grammar of astrology, and tables for calculating nativities, by Zadkiel. London, H. G. Bohn, 1852. xiv, 492, 64 pp., illus., plates.

1369 d

Lindley, Sir J. The elements of botany, structural, physiological, & medical: being a sixth edition of the Outline of the first principles of botany . . . new edition, with some corrections. London, Bradbury & Evans, 1849. xii, 142, c pp., illus. [*see* 1370].

1370 a d h

— The vegetable kingdom: or, the structure, classification, and uses of plants . . . third edition, with corrections and additional genera. London, Bradbury & Evans, 1853. lxviii, 908 pp., front., illus.

1371

Literary and Philosophical Society of Liverpool. Suggestions offered on the part of the . . . Society . . . to members of the mercantile marine . . . for the promotion of science, in furtherance of zoology. Liverpool, T. Brakell, printer for the Society, 1862. 51 pp. [*see* 228].

1372 d

Long, C. A. Practical photography, or glass and paper; fourth edition. London, Bland, 1859. 77 pp., illus. [*see* 223].

1373 a d h

Lyell, Sir C. Principles of geology: or the modern changes of the earth and its inhabitants . . . eleventh and entirely revised edition. London, John Murray, 1872. 2 vols., fronts., illus., plates.

1374 l

Maitland, A. C. What shall we have for breakfast? or everybody's breakfast book. London, John Hogg, 1889. 120 pp.

1375 j

Mann, J. A. Cocoa: its cultivation, manufacture, and uses . . . London, W. Trounce, 1860. 24 pp. (J. A. Mann, London, 24 Nov. 1862 to Dr Watson). [*see* 222].

1376 k

Massoul, C. de. A treatise on the art of painting, and the composition of colours, containing instructions for all the various processes of painting; translated from the French. London, T. Baylis, printer, 1797. [iv], 242 pp.

1377 d

Miers, E. J. On a small collection of crustacea by Major [*sic*] Burton in the Gulf of Akaba. [London, 1878]. 406–11 pp. Reprinted from the *Annals and magazine of natural history,* Nov. 1878. [*see* 64].

1378 a d

Mitchell, J. Manual of practical assaying, intended for the use of metallurgists, captains of mines, and assayers in general; second edition, entirely revised and greatly enlarged. London, New York, Hippolyte Baillière, [etc.], 1854. xi, 568 pp., illus.

1379 a d f h

Norrie, J. W. A complete epitome of practical navigation . . . to which is added a correct and extensive set of tables; seventeenth (stereotype) edition, considerably augmented and improved . . . by George Coleman . . . London, printed for the author, and sold by Charles Wilson, [1860]. xii, 344, xliv, 360 (of tables) pp., front., illus., plates, maps.

1380
Norton, J. L. Norton's patent tube well, as used by the army in Abyssinia. London, J. L. Norton, [1869]. 46 pp., illus. [*see* 228].

1381
Odell, G., printer. Odell's system of short-hand; forty-eighth edition. London, Groombridge, [?1850]. 16 pp., plates. [*see* 223].

1382 a d
Ottoni, C. B. Elementos de algebra: compendio adoptado pelos estabelecimentos de instrucção superior e secundaria; segunda edição com additamentos e numerosas correcções. Rio de Janeiro, á venda em casa de Eduardo & Henrique Laemmert, 1856. v, 210 pp.

1383 a d h
— Elementos de arithmetica; compendio adoptado pelos estabelecimentos de instrucção superior e secundaria; quarta edição. Rio de Janeiro, publicado e á venda em casa dos editores Eduardo & Henrique Laemmert, 1864. [v], 222 pp.

1384 a d
— Elementos de geometria e trigonometria rectilinea: compendio adoptado por todos os estabelecimentos de instrucção secundaria e superior; segunda edição. Rio de Janeiro, publicado e à venda em casa dos editores Eduardo & Henrique Laemmert, 1857. vi, 214, 46 pp., plates.

1385 d
Page, D. Rudiments of geology. Edinburgh, William & Robert Chambers, 1851. 222 pp., illus., maps. [*see* 1327].

1386
Penn, G. Conversations on geology: comprising a familiar explanation of the Huttonian and Wernerian systems; the mosaic geology. London, Samuel Maunder, 1828. xxiv, 371 pp., illus., plates.

1387
Philippi, R. A. Elementos de historia natural. Santiago, Imprenta i Libreria de la Independencia, 1866. 328 pp.

1388 k
Phillips, W. A selection of facts from the best authorities, arranged so as to form an outline of the geology of England and Wales. London, William Phillips, printer, 1818. [ii], 240, [10] pp., front., illus., map.

1389 k
Plattner, C. F. Plattner's manual of qualitative and quantitative analysis with the blowpipe; from the last German edition, revised and enlarged by Th. Richter, translated by Henry B. Cornwall, assisted by John H. Caswell; second edition revised. New York, D. van Nostrand, 1873. xvii, 549 pp., illus., plate.

1390 a d
Pokorny, A. Storia illustrata del regno vegetale; versione italiana di Teodoro Caruel. Roma, Torino, Firenze, Ermanno Loescher, 1871. viii, 208 pp., illus.

1391
Qānūn al-masāhāt wa al-akīyāl wa al-auzān al-jadīdah [Tables of weights and measures; Arabic text]. [n.p.; n.d.]. 57 pp. [*see* p.].

1392 d
Raper, H. The practice of navigation and nautical astronomy; sixth edition. London; J. D. Potter, 1857. xxvi, 908 pp., illus.

1393
Riddell, R. Indian domestic economy and receipt book; seventh edition revised. Calcutta, Thacker, Spink; Bombay, Thacker, Vining; London, Thacker, 1871. ix, 633 pp., tables.

1394
Ridge, J. J. Diet for the sick: being nutritious combinations suitable for severe cases of illness; third edition. London, J. & A. Churchill, 1886. 56 pp.

1395 i
Ross, W. A. Pyrology, or fire chemistry: a science interesting to the general philosopher, and an art of infinite importance to the chemist, mineralogist . . . &c., &c. London, New York, E. & F. N. Spon, 1875. xxviii, 346 pp., front., illus., plates. (14 Sept. 1878).

1396 l
Salis, Mrs H. A. de. Dressed game and poultry à la mode. London, New York,

Longmans, Green, 1888. [vi], 79 pp.

1397 d

Scrope, G. P. Volcanos: the character of their phenomena, their share in the structure and composition of the surface of the globe . . . second edition, revised and enlarged. London, Longman, Green, [etc.], 1862. xii, 490 pp., front., illus.

1398 e

Silk Supply Association. Report of the proceedings of a meeting held in London, February 18, 1868. London, 1869. 16 pp. [*see* 256].

1399 i

Silva Maia, E. J. da. Quadros synopticos do reino animal. Rio de Janeiro, Typographia Nacional, 1858. xiii pp., tables. (15 Nov. 1865). [*see* 250].

1400

Smith, G. The laboratory: or, school of arts . . . compiled from German, and other foreign authors; third edition. London, James Hodges & T. Astley, 1750–6. 2 vols., fronts., illus., plates.

1401

Spigel, A. Adriani Spigelii, Philos. ac Medici Patauini Isagoges in rem herbariam; libri duo. Lugduni Batavorum, ex officina Elzeviriana, 1633. 272, [16] pp.

1402 a d

Steinbuechel de Rheinwall, A. Il commercio in grande e le stazioni marittime. Trieste, Tip. del Lloyd Austriaco, 1869. 18 pp. Reprinted from *Osservatore triestino,* Dec. 1868. [*see* p.].

1403 d

Stossich, A. Sostanze alimentari delle plante. [Trieste, L. Herrmanstorfer, 18—]. [ii], 6 pp. [*see* p.].

1404 a d

Wells, W. C. An essay on dew and several appearances connected with it; edited, with annotations by L. P. Casella, and an appendix by R. Strachan. London, Longmans, Green, [etc.], 1866. xi, 152 pp.

1405 k

White, J. A compendious dictionary of the veterinary art; also, a short description of the anatomy or structure of the eye, the foot, and other important parts of the horse; with practical observations on his diseases; second edition, greatly enlarged. London, Longman, Rees, Orme, [etc.], 1830. [498] pp., front.

1406 h

Wigzell, Halsey & Co. Greek-fire torpedo-boats. London, Smart & Allen, [1877]. [ii], 32 pp., plates. [*see* p.].

1407

Ẓāhir Khair Allāh, al-Shuwairī. Kitāb tadhkirat al-Kuttāb fi 'ilm al-hisāb. [A work on mathematics; Arabic text. Beirut, Matba'at al-Watanīyah, 1868]. 224 pp.

1408–88 Sword

1408
Anonymous. Leitfaden zum Bajonettiren. [Potsdam, Decker, printer, 18—]. 18 pp. [*see* 237].

1409 d
— Le trésor de la Maison impériale d'Autriche au Palais I. R. de la Bourg à Vienne. [Vienna, 18—]. 148 pp.

1410
Abd el Kader, L'Emir el Hadj. Règlements militaires . . . texte et traduction nouvelle accompagnée de notes par F. Patorni. Alger, Imprimerie P. Fontana, 1889. 94 pp. Date on wrapper 1890. [*see* p.].

1411 a d j
[Adjutant-General's Office]. Infantry sword exercise; revised edition; Adjutant-General's Office, Horse Guards. London, H.M. Stationery Office, [1842]. 40 pp. front. imperfect. (H. Schültz Wilson, [London], 30 Oct. 1875). [*see* 107, Box 4].

1412
André, E. ed. Le jeu de l'épée: leçons de Jules Jacob, rédigées par Emile André suivies du duel au sabre et du duel au pistolet . . . troisième édition. Paris, Paul Ollendorff, 1887. xxxvi, 278 pp.

1413 i
[Angelo, H. C.]. Bayonet exercise. London, Parker, Furnivall & Parker, [1853]. 30 pp., plates. [*see* 236].

1414 a h j
Arista, S. M. Del progresso della scherma in Italia . . . Bologna, Società tipografica già Compositori, 1884. 32 pp. (Walter Gregor, Fraserburgh, 25 Dec. 1884). [*see* p.].

1415 d f
Bayerische Nationalmuseum. Das bayerische Nationalmuseum. München, Kgl. Hofbuchdruckerei von Dr C. Wolf & Sohn, 1868. vii, 381 pp., illus., plate.

1416 a d h
Boissier, G. Le Musée de Saint-Germain, Musée des Antiquites nationales. Paris, se vend au profit de l'Hopital-Hospice, 1882. 72 pp., illus. Reprinted from *Revue des deux-mondes.* [*see* 107, Box 3].

1417 a d
British Museum. Catalogue of a series of photographs . . . from the collection . . . The catalogue of the VII. parts are by A. W. Franks, S. Birch, Geo. Smith, and Walter de Gray Birch. London, W. A. Mansell, [1872]. lii, 122 pp.

1418 a d f
Browne, S. B. A practical guide to squad and setting-up drill . . . Part I, Field exercise of the Army; second edition. London, Wm. H. Allen, 1871. 84 pp., illus.

1419 a d
Capellini, C. G. Armi e utensili di pietra del Bolognese descritti e figurati. Bologna, Tipi Gamberini e Parmeggiana, 1870. 16 pp., plate. Reprinted from *Memorie dell'Accademia delle Scienze dell'Istituto di Bologna,* Series 2, Vol. 9. [*see* pq.].

1420 h
Carl von Preussen, Prinz, collector. Waffen-Sammlung Sr. Königlichen Hoheit des Prinzen Carl von Preussen. Mittelalterliche Abtheilung; beschrieben und zusammengestellt sowie mit historischen Bemerkungen und Erläuterungen versehen von Georg Hiltl. Berlin, W. Moeser, [1876]. vi, 195 pp., fronts., illus. [*see* q.].

1421
Castle, E. Schools and masters of fence from the middle ages to the eighteenth century . . . London, George Bell, 1885. lii, 254 pp., front., illus., plates.
1422 d i
Chapman, G. Foil practice; with a review of the art of fencing . . . London, W. Clowes, 1861. 50 pp., plates. (June 1875).
1423 d
— Notes and observations on the art of fencing: a sequel to 'Foil practice' . . . Part 1, No. 1. London, Clowes, 1864. [ii], 18 pp., plates. [*see* 1422].
1424
Chatin, M. Escrime à la bayonnette. Paris, de Blot, 1856. viii, 48 pp., illus. [*see* 236]. Another copy [*see* 232].
1425 a d i
Coustard de Massi, A. P. History of duelling in all countries; translated from the French . . . with introduction and concluding chapter by Sir Lucius O'Trigger. London, Newman, [1880]. xxxix, 116 pp.
1426
Demmin, A. Guide des amateurs d'armes et armures anciennces par ordre chronologique depuis les temps les plus reculés jusqu'à nos jours; deuxième édition. Paris, Renouard, Henry Loones, 1879. [iii], 628 pp., illus. At head of title: Encyclopédie d'armurerie avec monogrammes.
1427 a d
— An illustrated history of arms and armour from the earliest period to the present time . . . translated by C. C. Black. London, George Bell, 1887. vii, 595 pp., imperfect.
1428 d f g
Dresden. Königliches historisches Museum. Führer durch das Königliche historische Museum im Zwingergebäude zu Dresden. Dresden, [18—]. 80 pp.
1429 d i
[Faulder, W. W., collector]. Exhibition of industrial art, etc., Ancoates, June and July, 1881. [Manchester, A. Ireland, printer], 1881. 32 pp.
1430
Fine Art Society, London, Catalogue of, and notes upon the loan exhibition of Japanese art held at the Fine Art Society's. London, 1888. 116 pp. At head of title: Exhibition No. 56.
1431
[France. Ministère de la Guerre]. Escrime à la baïonnette: extraits de l'ordonnance du 22 juillet 1845 sur l'exercice et les manoeuvres des bataillons des chasseurs a pied . . . Paris, J. Dumaine, 1852. 20 pp., plates. Date on wrapper 1853.
1432 d l
— — Escrime à la bayonnette suivie du supplément à l'Ecole des Tirailleurs, pour la formation des carrés par quatre et par huit, en usage dans les Bataillons de Chasseurs à Pied (Ordonnance du 22 juillet 1845) . . . Paris, Imprimerie Lith. et Librairie Militaires de Blot, 1852. 12 pp., illus. [*see* 54].
1433 a
— — Règlement du 12 juin 1875 sur les manoeuvres de l'infanterie . . . suivi des Exercices d'assouplissement. Paris, J. Dumaine, 1875. lxx, 148 pp., *and* 23 pp.
1434 a d f g
— — Règlement provisoire sur les exercises de la cavalerie. Tome premier . . . 3e tirage. Paris, J. Dumaine, 1873. xx, 318 pp., illus., plates.
1435 a d
— Ministère de la Marine et des Colonies. Manuel pour l'enseignement de la gymnastique et de l'escrime . . . Paris, J. Dumaine, 1875. [iv], 258 pp., front., illus.
1436 i
Graesse, J. G. T. Guide de l'amateur d'objets d'art et de curiosité ou collection des monogrammes des principaux sclupteurs en pierre, métal et bois . . . et des médailleurs . . . deuxième édition. Dresde, G. Schoenfeld, 1877. (A. Hosty).
Another copy.
1437 d
Great Britain. Army. Adjutant General's Office. Infantry sword exercise; revised edition. London, Parker, Furnivall & Parker, 1845. 46 pp., plate. [*see* 236].
1438
— — — Instructions for the carbine exercise: the pistol exercise; and the lance exercise; revised and corrected. London, Parker, Furnivall & Parker, 1850. 35 pp. [*see* 236].

1439 a
— — — Instructions for the sword, carbine, pistol, and lance exercise, together with field gun drill for the use of the cavalry. London, Her Majesty's Stationery Office, [1871]. vi, 144 pp., diagrs.

1440 a d h
Grisier, A. Les armes et le duel . . . préface anecdotique par Alexandre Dumas; troisième édition. Paris, E. Dentu, 1864. 607 pp., front., plates.

1441
[Habelmann, — and others]. Praktische Anleitung zum unterricht im Stossfechten. Berlin, E. H. Schroeder, Hermann Kaiser, 1872. 60 pp., illus.

1442
Haddan, J. L. Iron-clads and forts: their armour, guns, propellers, turrets, &c. London, Edward Stanford, 1872. [iii], 70 pp., front., plates. At head of title: For private circulation only. [*see* p.].

1443
Harman, A. Sketch of the Tower of London, as a fortress, a prison, and a palace . . . also a guide to the armories. London, J. Wheeler, [1872]. iv, 44 pp., front., illus., plates. [*see* p.].

1444 a d
[Harris, W. C.]. A manual of drill and sword exercise, prepared for the use of the county and district constables; fourth edition. London, W. Clowes, 1868. viii, 120 pp., front., illus., plates.

1445 a e
Hergsell, G. Die Fechtkunst. Wien, Pest, Leipzig, A. Hartleben, 1881. xiv, 358 pp., front., plates.

1446 d f h
Innsbruck. Ferdinandeum. Die Sammlungen im Landes-Museum (Ferdinandeum) zu Innsbruck. Innsbruck, Wagner'sche Univeristäts-Buchdruckerei, 1878. 92 pp.

1447
[Italy. Ministero della Guerra]. Istruzione per la Scuola di Scherma, colla baionetta; approvata dal Ministero della Guerra (maggio 1856). Torino, dall'Officina Tipografico dei Fratelli Fodrattai, [*c.* 1856]. 27 pp. [*see* 54].

1448 a d
La Boëssière, —. Traité de l'art des armes, à l'usage des professeurs et des amateurs. Paris, Imprimerie de Didot, l'Ainé, 1818. xxii, 309 pp., plates.

1449 e h
Lacombe, P. Les armes et les armures; deuxième édition. Paris, Hachette, 1870. [iii], 302 pp. (incl. plates), illus., plates.

1450 d
Lane Fox, A. H. Primitive warfare [section 1]: illustrated by specimens from the Museum of the [Royal United Service] Institution. [London, Harrison, printer], 1867. 35 pp., plates. [*see* 241].

1451 d
— — section II: on the resemblance of weapons of early races; their variations, continuity, and development of form . . . a lecture delivered at the Royal United Service Institution. [London, Harrison, printer], 1868. 42 pp., plates. At head of p. [2]: (For private circulation only). [*see* 241].

1452 d
Latham, J. Arms, and the men. [London, 186-?]. 223–30 pp. Reprinted from an unidentified journal. [*see* 107, Box 4].

1453 d f i
— A few notes on the swords in the International Exhibition. [London, 1863]. 11 pp. Reprinted from the *Journal of the Royal United Service Institution,* Vol. 7. [*see* 107, Box 4].

1454 a d h
— The shape of sword-blades. [London], 1862. 13 pp. Reprinted from the *Journal of the Royal United Service Institution,* Vol. 6 [*see* 107, Box 4].

1455 a d f
Luebeck, W. Lehr- und Handbuch der deutschen Fechtkunst; zweite Ausgabe. Frankfurt a.d. Oder, Gustav Harnecker, 1869. ix, 274 pp., plates.

1456 d i l
McClellan, G. B. Manuel of bayonet exercise; prepared for the use of the Army of the United States. Philadelphia, Lippincott, Grambo, 1852. xii, 106 pp., plates. (J. J. Dana, U.S. 4th Arty.).

1457 d j l

Maclaren, A. A system of fencing, for the use of instructors in the Army. London, Her Majesty's Stationery Office, [1864]. viii, 80 pp., illus. (A. MacLaren, Oxford, Sunday Evening; Oxford, 22 July 1876).

1458 h

Madrid. Real Armería. Catálogo de la Real Armería, mandado por . . . José María Marchesi. Madrid, Aguado, 1849. xx, 198, 120 pp., plates.

1459 a d f

Marchionni, A. Trattato di scherma sopra un nuovo sistema di Giuoco Misto di Scuola italiana e francese . . . [Fascicolo 1]. Firenze, Tipi di Federigo Bencini, 1847. vii, 376 pp., illus., plates.

Another copy. imperfect.

1460 a d

Marozzo, A. Arte dell'armi . . . ricorretto, et ornato di nuove figure in rame. In Venetia, appresso Antonio Pinargenti, 1568. [xvi], 194 pp. (pagination irregular). imperfect.

1461

Mérignac, É. Histoire de l'escrime dans tous les temps et dans tous les pays . . . II, Moyen âge-temps modernes. Paris, Rouquette, 1886. [v], 598 pp., illus., plates.

1462 d

Millotte, C. D. F. Traité d'escrime, pointe. Paris, J. Dumaine, 1864. 65 pp., illus.

1463

Moncrieff, A. Moncrieff's protected barbette system: from the "Proceedings" of the R. A. Institution, Woolwich. Woolwich, Royal Artillery Institution, 1868. 14 pp., illus. [*see* 222].

1464

— On the Moncrieff system of working artillery as applied to coast defence. London, Royal Institution of Great Britain, 1869. 16 pp. [*see* 222].

1465

Munich. Schützenordnung. Schützen-Ordnung [für die Churfürste. Haupt- und Residenzstadt, und sämmtliche Schiess-Stätte der Churpfalzbaierischen Lande]. München, Franz Seraph Hübschmann, printer, 1796. 54 pp.

1466 d

Notes and queries: a medium of intercommunication for literary men, general readers, etc. Series 5, Vol. 4, Nos. 89, 91–2, 94, 96. London, 1875. See 'A list of works on sword play', by Fred. W. Foster. [*see* p.].

1467 a d

Oppert, G. On the weapons, army organisation, and political maxims of the ancient Hindus, with special reference to gunpowder and firearms. Madras, Higginbotham; London, Trübner, 1880. vii, 162 pp., tables.

1468 a d f

Paris. Musée des Thermes et de l'Hotel de Cluny. Catalogue et description des objets d'art de l'antiquité du Moyen âge et de la Renaissance exposés au Musée, par E. du Sommerard. Paris, Hotel de Cluny, 1881. xxxiii, 690 pp.

1469 a d f

Penguilly l'Haridon, O. Catalogue des collections composant le Musée d'Artillerie. Paris, Charles de Mourgues, 1862. xii, 1004 pp.

1470

Phillips, J. C., compiler. Abolition of the bonus system in the Indian army. London, Wm. H. Allen, 1869. 24 pp. [*see* p.].

1471 a h

Pilla, C. Arte e scuole de scherma: conferenza tenuta alla Società bolognese di Scherma nel Febbraio 1886. Bologna, Società Tipografica già Compositori, 1886. 46 pp.

1472 a d j l

Possellier, A. J. J. La théorie de l'escrime enseignée par une méthode simple basée sur l'observation de la nature . . . par A.-J.-J. Possellier dit Gomard. Paris, J. Dumaine, 1845. [vii], 323 pp., plates, tables, bibliogr. (postcard signed in Arabic: [R. S. Charnock], date stamped: South Kensington, 22 July 1884).

1473 i

[Rathborne, A. B.]. Opinions of counsel on the legality of the late Indian army bonus system. London, printed for private circulation, 1869. 16 pp. (Lt. Col. J. C. Phillips). [*see* p.].

1474 h j l

Schirlitz, F. L. ed. Atlas zur Fabrikation der Stahl-Waaren . . . in Solingen . . . [Weimar, Bernhard Friedrich Voigt, printer, after 1866]. 4 pp., plates. (J. H. Angst, Zürich, 9 May 1889; H. Ellen Bishop, Hastings, 13 July 1889 – to Lady Burton; postcard signed in Arabic: [R. S. Charnock], date stamped New Thornton Heath, 4 July 1884). [*see* p.].

1475 a

Shirley, A. Remarks on the transport of cavalry and artillery . . . London, Parker, Furnivall & Parker, 1864. 33 pp. [*see* 236].

1476 a d

South Kensington Museum. Bethnal Green Branch Museum. Catalogue of the anthropological collection lent by Colonel Lane Fox for exhibition . . . June 1874. Parts I and II. London, Her Majesty's Stationery Office, 1874. xvi, 184 pp., plates, diagr.

1477

Tennent, Sir J. E. The story of the guns. London, Longman, Green, 1864. xxii, 364 pp., front., illus., plates.

1478 a d h j

Truman, B. C. The field of honor: being a complete and comprehensive history of duelling in all countries . . . New York, Fords, Howard & Hulbert, 1884. 599 pp. (postcard signed in Arabic: [R. S. Charnock], date stamped: London, 5 May 1885).

1479

United States Army. Official army register, for 1860. Washington, Adjutant General's Office, 1860. 60 pp. [*see* 244].

1480

United States Military Academy. Official register of the officers and cadets of the U.S. Military Academy, West Point, New York, June 1859. [New York, John F. Baldwin, printer], 1859. 20 pp. [*see* 244].

1481

— Regulations for the U.S. Military Academy at West Point, New York. New York, John F. Trow, printer, 1857. 172 pp. [*see* 245].

1481.1 l

Urbani de Gheltof, D. Difesa di un vecchio pugnale veneziano. [Venezia], 1876. 23 pp., illus. [*see* 107, Box 2].

1482 d f h

[Vienna]. Waffensammlung. Catalogue de la collection I. & R. d'Ambras. Vienne, 1870. iv, 40 pp. Title and date on wrapper: Notice sommaire de la collection . . . 1873.

1483 d

— Catalogue du Musée d'Armes de la Cour impériale; traduit de l'Allemand par le Major M. Prosig. Vienne, 1870. viii, 52 pp.

1484 a d f

— Katalog des Waffenmuseums der kaiserl. Haupt- und Residenzstadt Wien; zweite verbesserte und erweiterte Auflage. Wien, Selbstverlag des Gemeinderathes, 1877. vii, 71 pp.

d

Another copy.

1485 h

Vigeant, F. La bibliographie de l'escrime ancienne et moderne . . . Paris, Motteroz, 1882. 172 pp. [*see* 107, Box 4].

1486 d

Waite, J. M. Lessons in sabre, singlestick, sabre & bayonet, and sword feats . . . London, Weldon, [1880]. 151 pp. wanting title-page and wrappers.

1487 a d

Wilkinson, H. Observations on swords addressed to officers and civilians . . . fifteenth edition. London, Pall Mall, [Wilkinson & Son; 187–]. 40 pp., illus.

1488 a

Worsaae, J. J. A. Nordiske oldsager i det Kongelige Museum i Kjöbenhavn, ordnede og forklarede. Kjöbenhavn, Kittendorff & Aagaard, 1859. 200 pp.

1489-711 Africa

1489 d
Anonymous. The Nile basin. London, [1865]. 18 pp. Review of *The Nile basin,* by R. F. Burton and James M'Queen. From *Colburn's new monthly magazine,* Jan. 1865, Vol. 133. [*see* 228].

1490 d
— Nile basins and Nile explorers. [London], 1865. 100–17 pp. From an unidentified journal, Jan. 1865. Review of *The Nile basin,* by R. F. Burton and James McQueen and other works. [*see* 220].

1491 l
— [Review of Abeokuta and the Camaroons mountains, and A mission to Gelele, king of Dahome, by R. F. Burton. 1864]. 452–83 pp. From an unidentified journal, Vol. 23, No. 46. [*see* 228].

1492
— [Review of The basin of the upper Nile and its inhabitants, by J. H. Speke and works by C. T. Beke and H. Barth]. London, 1864. 307–48 pp. From the *Westminster and foreign quarterly review,* Vol. 81. [*see* 241].

1493 d
— [Review of]. Voyage au Congo et dans l'interieur de l'Afrique equinoxiale, fait dans les années 1828, 1829 et 1830, par J. B. Douville. London, 1832. 163–206 pp. From the *Foreign quarterly review,* Vol. 10, No. 19. [*see* 218].

1494 d i
Abbate, O. De l'Afrique centrale: ou voyage de S. A. Mohammed-saïd-Pacha dans ses provinces du Soudan. Paris, Henri Plon, 1858. [i], 54 pp. (Cairo, June 1876).

1495 d l
Alexander, Sir J. E. Narrative of a voyage of observation among the colonies of Western Africa . . . and of a campaign in Kaffir-land . . . in 1835. London, Henry Colborn, 1837. Vol. 1, pp. 113–378, plates, tables, map. wanting title-page, and pp. 337–48. [*see* 238].

1496
[Allemand-Lavigerie, C. M.]. L'esclavage africain: conférence fait à Saint-Sulpice. [Saint Cloud, Imprimerie Ve Eugene Belin, 1888]. 52 pp. [*see* p.].

1497 a
— L'esclavage africain: conférence sur l'esclavage dans le haut Congo faite à Sainte-Gudule de Bruxelles. Bruxelles, Société Anti-Esclavagiste; Paris, procure des Missions d'Afrique, 1888. 35 pp. [*see* p.].

1498 a
— L'esclavage africain: discours prononcé au meeting tenu à Londres, le 31 juillet 1888 sous la présidence de Lord Granville. Paris, procure des Missions d'Afrique, 1888. 27 pp. [*see* p.].

1499 d
Allen, W. A narrative of the expedition sent by Her Majesty's Government to the river Niger, in 1841, under the command of Captain H. D. Trotter, R.N. . . . and T. R. H. Thomson. London, Richard Bentley, 1848. 2 vols, fronts., illus., plates, maps.

1500 d
Alison, R. E. Teneriffe: an ascent of the peak and sketch of the island. [London], 1866. iii, 24 pp., front. From the *Quarterly journal of science,* Vol. 9. [*see* 240].

1501 a d g h
Amicis, E. de. Morocco: its people and places; translated by C. Rollin-Tilton. London,

Paris, New York, Cassell, Petter, Galpin, 1882. 392 pp.

1502 i
Artin, Y. L'instruction publique en Egypte. Paris, Ernest Leroux, 1889. viii, 214 pp., tables.

1503 i
— La propriété foncière en Egypte. Le Caire, Imprimerie Nationale de Boulaq, 1883. 349 pp., tables. At head of title: Institut égyptien.

1504 d
— Les trois femmes et le Kadi – Bab Zoueyleh et la mosquée d'El-Moéyed . . . malice des femmes. Le Caire, J. Barbier, 1884. 68 pp. Reprinted from *Bulletin de l'Institut égyptien,* Series 2, Vol. 4. [*see* p.].

1505 k
Aubertin, J. J. Six months in Cape Colony and Natal, and one month in Tenerife and Madeira. London, Kegan Paul, Trench, 1886. [vii], 279 pp., front., plates, map.

1506
Baedeker, K. Aegypten: Handbuch für Reisende. Erster Theil: Unter-Aegypten bis zum Fayûm und die Sinai-Halbinsel. Leipzig, Karl Baedeker, 1877. xvi, 562 pp., front., illus., plates, maps.

1507 d
Baker, Sir S. W. The Albert N'yanza, great basin of the Nile, and explorations of the Nile sources. London, Macmillan, 1866. 2 vols., fronts., illus., plates, maps.

1508
— The Egyptian question: being letters to The Times and Pall Mall gazette. London, Macmillan, 1884. [iii], 94 pp., front.

1509
Barth, H. Travels and discoveries in north and central Africa: being a journal of an expedition undertaken . . . in the years 1849-1855. London, Longman, Brown, Green, [etc.], 1857-8. 5 vols., fronts., illus., plates, maps.

1510 l
Beaufort, E. A. Egyptian sepulchres and Syrian shrines, including some stay in the Lebanon at Palmyra, and in western Turkey. London, Longman, Green, [etc.], 1861. 2 vols., fronts., illus., plates, maps.

1511 a d
Beke, C. T. The British captives in Abyssinia. London, Longman, Green, [etc.], 1865. [i], 61 pp. [*see* 241].

1512 d i
— The French and English in the Red Sea; second edition. London, Taylor & Francis, 1862. 30 pp. [*see* 241].

1513 i
— A lecture on the sources of the Nile, and on the means requisite for their final determination; delivered in the theatre of the London Institution . . . London, Board of Management of the London Institution, 1864. 35 pp., front., maps.

i
Another copy. [*see* 241].

1514 d i
— On the mountains forming the eastern side of the basin of the Nile [read before the British Association for the Advancement of Science. London], 1861. 15 pp. [*see* 238].

1515 a d
— The sources of the Nile: being a general survey of the basin of that river, and of its head-streams; with the history of Nilotic discovery. London, James Madden, 1860. xix, 156 pp., front., plate, maps, tables.

1516 i
— Who discovered the sources of the Nile? a letter to Sir Roderick I. Murchison; with an appendix; second edition. London, Williams & Norgate, 1863. [ii], 16 pp. [*see* 241].

1517
Belcastel, G. de. Las Islas Canarias e el valle de Orotava; version literal por Aurelio Perez Zamora. Santa Cruz de Tenerife, Imprenta y Litografia Islenã, 1862. 38 pp. [*see* 241].

1518
Bonelli, E. El Imperio de Marruecos y su constitucion. Madrid, Imprenta y Litografía del Deposito de la Guerra, 1882. [iii], 270 pp.

1519 d
[Borelli, J.]. Divisions, subdivisions, langues et races des régions Amhara, Oromo et Sidama [Annexes B–F]. [Paris, Libraries-Imprimeries Réunies, 1890]. 68 pp. From

Ethiopie méridionale: journal de mon voyage aux pays Amhara, [etc.], by J. Borelli, pp. 431–98. [*see* pq.].

1520 a d f

Boteler, T. Narrative of a voyage of discovery to Africa and Arabia, performed in His Majesty's ships Leven and Barracouta, from 1821 to 1826, under the command of Capt. F. W. Owen, R.N. London, Richard Bentley, 1835. 2 vols., fronts., illus., plates.

1521 a d

Bowdich, T. T. Mission from Cape coast castle to Ashantee, with a statistical account of that kingdom, and geographical notices of other parts of the interior of Africa. London, John Murray, 1819. x, 512 pp., front. (map), plates, map.

1522 d i

Brazza, P. Savorgnan de. Réception de Monsieur P. Savorgnan de Brazza enseigne de vaisseau au grand amphithéatre de la Sorbonne le 23 juin 1882. Paris, Société de Géographie, 1882. 277–300 pp., map. Reprinted from *Compte-rendu des séances de la Société de Géographie,* No. 13. [*see* p.].

1523 a d

Broadley, A. M. How we defended Arábi and his friends: a story of Egypt and the Egyptians; illustrated by Frederick Villiers. London, Chapman & Hall, 1884. xiii, 507 pp., front. (port.), plates.

1524 d j

Browne, A. J. J. On some flint implements from Egypt. [London], 1878. 17 pp., plate, map. Reprinted from the *Journal of the Anthropological Institute,* Vol. 7. (W. E. Haynes, of Greenfield & Co., Alexandria Harbour Contract, Alexandria, 16 Sept. and 10 Oct. 1878). [*see* p.].

1525 a d l

Browne, W. G. Travels in Africa, Egypt, and Syria from the year 1792 to 1798; second edition, enlarged. London, printed for T. Cadell & W. Davies, 1806. [3], xxxv, 632 pp., front., illus., maps.

1526

Bruce of Kinnaird, J. Travels through part of Africa, Syria, Egypt, and Arabia into Abyssinia, to discover the source of the Nile. Halifax, William Milner, 1846. xvi, 463 pp., front., plates.

1527 a d h j

Brugsch, H. Geschichte Aegypten's under den Pharaonen, nach den Denkmälern; erste deutsche Ausgabe. Leipzig, J. C. Hinrichs, 1877. xiii, 818 pp., front. (map), illus., map. (H. Brugsch, Goettingen, 9 Oct. 1877).

1528 a d

— Histoire d'Egypte. Première partie, Introduction: Histoire des dynasties I–XVII; deuxième édition. Leipzig, J. C. Hinrichs, 1875. [iii], 180 pp., illus.

1529 d l

Bryson, A. Report on the climate and principal diseases of the African station; compiled from documents in the Office of the Director-General of the Medical Department . . . London, William Clowes, printer, 1847. xv, 266 pp., tables.

1530 d h

Budge, Sir E. A. T. W. The dwellers on the Nile: or chapters on the life, literature, history and customs of the ancient Egyptians. [London], Religious Tract Society, 1885. 204 pp., front., illus. (By-paths of Bible knowledge, Vol. 8).

1531 a d g h

Bunsen, C. C. J. Egypt's place in universal history: an historical investigation in five books; translated from the German by Charles H. Cottrell; with additions by Samuel Birch. Vol. 5. London, Longmans, Green, 1867. xviii, 943 pp., tables.

1532 a d

Burckhardt, J. L. Arabic proverbs: or the manners and customs of the modern Egyptians, illustrated from their proverbial sayings current at Cairo; translated and explained; second edition. London, Bernard Quaritch, 1875. viii, 283 pp., illus.
Another copy.

1533

Burton, R. pseud. of Nathaniel Crouch. The English acquisitions in Guinea and East-India, containing, first, the several forts and castles of the Royal African Company from Sally in south Barbary, to the Cape of Good-Hope, in Africa . . . secondly, the forts and factories of the Honourable East-India Company in Persia, India, Sumatra, China &c. . . . London, printed for A. Bettesworth [etc.], and J. Batley, 1728. 184 pp.

1534 a d h i

Cameron, V. L. Across Africa. London, Daldy, Isbister, 1877. 2 vols., fronts., illus., plates.

1535
— The Congo and other river basins of Africa clearly set out. London, George Berridge, [1884]. [4] pp. At head of title: Berlin Conference. [*see* p.].

1536 a d h
Chaillé Long, C. Address on Egypt, Africa and Africans, delivered before the American Geographical Society . . . 1878. [n.p.], printed for the author, 1878. 39 pp. [*see* p.].

1537
Chesson, F. W. The war in Zululand: a brief review of Sir Bartle Frere's policy, drawn from official documents. [London], T. S. King, 1879. 26 pp. [*see* p.].

1538 d i
Clarke, R. Remarks on the topography and diseases of the Gold Coast. [London, Epidemiological Society], 1860. 54 pp., plate, maps. (Robert Clarke, London, 17 July 1861). [*see* 238].

1539 d
Cole, R. E. The river Gambia; read at York, September, 1881. Sydenham, Penge, Millard, printer, [1881]. 16 pp. [*see* p.].

1540 l
Cole, W. Life in the Niger: or, the journal of an African trader. London, Saunders, Otley, 1862. [iii], 208 pp.

1541
Comboni, D. Quadro storico delle scoperte africane. Verona, 1880. 94 pp. Reprinted from *Annali del Buon Pastore,* Vol. 22. [*see* p.].
Another copy.

1542 a d f
Compagnie Universelle du Canal Maritime de Suez. Assemblée générale des Actionnaires (. . . 2 juin 1868): rapport de M. Ferdinand de Lesseps. [Paris, Henri Plon], 1868. 80 pp., map, tables. [*see* 222].

1543 d
Cooley, W. D. Dr. Livingstone's errors. London, [186-]. 96–110 pp. From an unidentified journal. [*see* 225].

1544 a d h
— Inner Africa laid open, in an attempt to trace the chief lines of communication across that continent south of the equator, with the routes to the Muropue and the Cazembe, Moenemoezi and Lake Nyassa . . . London, Longman, Brown, Green, [etc.], 1852. viii, 149 pp., front. (map), table.

1545 a l
— The memoir on the lake regions of east Africa, reviewed: in reply to Capt. R. Burton's letter in the "Athenaeum", No. 1899. London, Edward Stanford, 1864. 16 pp. [*see* 241].

1546
Cordeiro, L. L'hydrographie africaine aux XVIe siècle d'après les premières explorations portugaises. Lisbonne, J. H. Verde, 1878. 72 pp. At head of title: Société de Géographie de Lisbonne. [*see* p.].

1547
— Viagens explorações e conquistas dos portuguezes collecção de documentos, 1593–1631, terras e minas africanas segundo Balthazar Rebello de Aragão. Lisboa, Imprensa Nacional, 1881. 24 pp. At head of title: Memorias do Ultramar. [*see* p.].

1548
Coveny, R. C. Letters . . . written during the campaigns in Egypt and the Soudan from 1882 to 1885. [n.p., n.d.], 39 pp., front.

1549 d
Cuny, C. Journal de voyage . . . de Siout à el-Obéid du 22 novembre 1857 au 5 avril 1858; precédé d'une introduction par M. V. A. Malte-Brun. Paris, Arthus Bertrand, 1863. 203 pp., map.

1550 a d
Dahse, P. The Gold Coast; translated from the German by Harry Bruce Walker. Liverpool, W. Barton, printer, 1882. 29 pp., map. [*see* p.].

1551 d
— Die Goldküste. Bremen, G. A. v. Halem, 1882. 81–110 pp., map. From *Deutsche geographische Blätter,* Vol. 5, No. 2. [*see* p.].

1552 a d h
Dalzel, A. The history of Dahomy: an inland kingdom of Africa; compiled from authentic memoirs. London, T. Spilsbury, printer for the author, 1793. xxx, xxxii, 230 pp., front. (map), plates.

1553 d l

Decken, C. C. von der. Reisen in Ost-Afrika in den Jahren 1859 bis 1861 [Vol. 2: 1862–1865]; bearbeitet von Orto Kersten, mit einem Vorworte von A. Petermann. Leipzig, Heidelberg, C. F. Winter, 1869–71. 2 vols. front. (port.), illus., plates, maps.

1554

Dennis, G. On recent excavations in the Greek cemeteries of the Cyrenaica. [London, 1870]. 48 pp., plates. Reprinted from the *Transactions of the Royal Society of Literature*, Series 2, Vol. 9. [*see* 241].

1555 d

Douville, J. B. Ma defense, ou reponse à l'anonyme anglais du *Foreign quarterly review*, sur le Voyage du Congo [and] Critical sketches. London, 1832. From the *Foreign quarterly review*, Vol. 10, No. 20, pp. 541–6, 547–52. [*see* 218]. *See also* 1556.

1556 a d

— Voyage au Congo et dans l'intérieur de l'Afrique equinoxiale, fait dans les années 1828, 1829 et 1830. Paris, chez Jules Renouard, 1832, 3 vols. *and* atlas, front., tables, plates.

1557 a

Drummond, H. Nyassaland: travel-sketches in our new Protectorate; selected from "Tropical Africa". London, Hodder & Stoughton, 1890. vii, 119 pp.

1558 a d h

Du Chaillu, P. B. Explorations & adventures in equatorial Africa. London, John Murray, 1861. xviii, 479 pp., illus., plates.

1559 a d i

— A journey to Ashango-land, and further penetration into equatorial Africa. London, John Murray, 1867. xxiv, 501 pp., illus., plates, map. (28 Jan. 1867).

1560 a d l

Duncan, J. Travels in western Africa in 1845 & 1846, comprising a journey from Whydah through the Kingdom of Dahomey, to Adofoodia. London, Richard Bentley, 1847. 2 vols. fronts., illus., plates, map.

1561 a d

Dupont, L. V. Marquis de Compiègne. Gabonais, Pahouins-Gallois. Paris, E. Plon, 1875. [v], 359 pp., front., plates, map.

1562 d

— Okanda, Bangouens-Osyéba. Paris, E. Plon, 1875. viii, 360 pp., front., plates, map. [*see* 1561].

1563 a d

Dupuis, J. Journal of a residence in Ashantee, comprising notes and researches relative to the Gold Coast, and the interior of western Africa; chiefly collected from Arabic MSS., and information communicated by the Moslems of Guinea. London, Henry Colburn, 1824. ix, xxxviii, 264, cxxxvi pp., front., plates, map.

1564 a d

East, D. T. Western Africa: its condition, and Christianity the means of its recovery. London, Houlston & Stoneman, 1844. xi, 400 pp. [*see* 1589].

1565

Egypt. Report on the Egyptian consolidated debt presented to H.H. the Khedive, by the Controllers-General and reply of H.H. the Khedive. Le Caire, Typographie française Delbos-Demouret, Jablin, 1880. 26 pp. [*see* p.].

1566 a d i

— Ministère de l'Intérieur. Statistique de l'Égypte, année 1873–1290 de l'Hégire. Le Caire, Imprimerie française Mourès, 1873. lxxxvi, 315 pp., tables. (Dr Paul Colucci, Alexandrie).

1567 a d

Egyptian calendar for the year 1295 A.H. (1878 A.D.) corresponding with the years 1594–1595 of the Koptic era. Alexandria, French Printing-Office, A. Mourès, 1877. 98 pp. Date on wrapper 1878. [*see* p.].

1568 a d

Ellis, A. B. West African islands. London, Chapman & Hall, 1885. viii, 352 pp.

1569 a d h

Ensor, F. S. Incidents on a journey through Nubia to Darfoor. London, W. H. Allen, 1881. viii, 227 pp., front. (map), map, tables.

1570 a i

Ferrand, G. Le Çomal. Alger, P. Fontana, 1884. 27 pp. Reprinted from the *Bulletin de correspondance africaine*, 1884, No. 4. [*see* p.].

1571*a*
Findlay, A. G. On Dr Livingstone's last journey, and the probable ultimate sources of the Nile. London, 1867. 16 pp. Reprinted from the *Journal of the Royal Geographical Society,* Vol. 37. Title on wrapper: Remarks on Dr. Livingstone's last journey in relation to the probable ultimate sources of the Nile. [*see* 241].
Another copy. [*see* 220].

1571*b* **g i**
Another copy entitled: Remarks [then as above]. [*see* p.].

1572 i
Fitzgerald, C. The Gambia, and its proposed cession to France. London, Unwin Brothers, 1875. 27 pp., front. (map). (9 Nov. 1875).

1573 h l
Flad, J. M. The Falashas (Jews) of Abyssinia, with a preface by Dr. Krapf; translated from the German by S. P. Goodhart. London, William Macintosh, 1869. xv, 75 pp., front.

1574 a d f h
Forbes, F. E. Dahomey and the Dahoomans: being the journals of missions to the King of Dahomey, and residence at his capital, in the years 1849 and 1850. London, Longman, Brown, Green, [etc.], 1851. 2 vols., fronts., plates.

1575 a d
Gamitto, A. C. P. O Muata Cazembe e os povos Maraves, Chévas, Muizas, Muembas, Lundas e outros da Africa austral: diario da expedição portugueza commandada pelo Major Monteiro, e dirigida aquelli Imperador nos annos de 1831 e 1832. Lisboa, Imprensa Nacional, 1854. xxv, 504 pp., front., plates, map, tables.

1576
[Garwood, A. E.]. Railway communication with central Africa: memorandum on Soudan railway. [Cairo, 1879]. 2 pp. [*see* p.].

1577 a d
Girard, A. Souvenirs d'un voyage en Abyssinie (1868-1869). Le Caire, Typographie française Delbot-Demouret, 1873. 312 pp.

1578
Gold Coast. Almanac for the year of Our Lord 1842. Cape Coast Castle, 1842. [61] pp. [*see* 239].

1579
[Gordon, C. G.]. Provinces of the Equator: summary of letters and reports of His Excellency the Governor-General. Part 1, year 1874. Cairo, Printing of the General Staff. 1877. viii, 90 pp., plate, maps. At head of title: Publications of the Egyptian General Staff. [*see* p.].

1580 a d h
Graham, A. and H. S. Ashbee. Travels in Tunisia. London, Dulau, 1887. viii, 295 pp., front., plates, map. (H. S. Ashbee, London, Xmas Day 1887; 6 Jan. 1888).

1581 a d l
Grant, J. A. A walk across Africa: or domestic scenes from my Nile journey. Edinburgh, London, William Blackwood, 1864. xviii, 453 pp., front. (port.), map (in pocket).

1582 d
[Great Britain. Admiralty. Hydrographic Office]. The African pilot: or sailing directions for the western coast of Africa. Part I, From Cape Spartel to the river Cameroons. Vol. 1. London, printed for the Hydrographic Office, Admiralty, 1856. viii, 238 pp., illus.
a
Another copy.

1583 a d i j
[Greene, J. B.]. The Hebrew migration from Egypt. London, Trübner, 1879. xi, 440 pp., front. (map), map. (J.B.G., post stamped London, 8 Nov. and 3 Dec. 1879; J. Baker Greene, [London], 20 Dec. 1879, 23 and 30 Jan. 1880; A. W. Thayer, 5 Oct. 1882).

1584 h
Guessfeld, P. Voyage du Dogteur [*sic*] Güssfeld à la côte occidentale d'Afrique [and other papers]. [Cairo, 1876]. 249-96 pp., front. (map), maps. From *Bulletin trimestriel de la Société khediviale de Géographie du Caire,* Series 1, Vol. 1/4, No. 3. [*see* p.].

1585 a d
Guillain, C. Documents sur l'histoire, la géographie et le commerce de l'Afrique orientale; recueillis et rédigés. Paris, Arthus Bertrand, [1856]. 3 vols. (in 2), tables, map.

1586 a e

Haddan, J. L. The Suez canal question viewed in a Cobdenian sense. Constantinople, 1873. 7 pp. [*see* p.].

1587 d

Hamilton, A. The river Niger and the progress of discovery and commerce in central Africa: a lecture. London, Dalton & Morgan, printer, 1862. 38 pp., map. [*see* 227].

1588

Harris, G. W. "The" practical guide to Algiers; fourth edition. London, George Philip, 1894. xviii, 176 pp., illus., plates, maps.

1589 a d

Hawthorne, N. ed. The journal of an African cruiser, comprising sketches of the Canaries, the Cape de Verds, Liberia, Madeira, Sierra Leone, and other places of interest on the west coast of Africa, by an officer of the U.S. Navy. Aberdeen, George Clark, 1848. 306 pp.

1590

Henry, R. C. Statement of the outrage committed upon an English factory in Benin river. Liverpool, Lee & Nightingale, printer, 1862. 18 pp. [*see* 241].

1591 a d

Herculano de Carvalho e Araujo, A. Roteiro da viagem de Vasco da Gama em MCCCVII; segunda edição, correcta e augmentada de algumas observações principalmente philologicas, por A. Herculano e o Barão do Castello de Paiva. Lisboa, Imprensa Nacional, 1861. xliii, 181 pp., front. (port.), illus., plate, map.

1592 d h

Hogg, J. On some old maps of Africa, in which the central equatorial lakes are laid down nearly in their true positions. [London, 1864]. 38 pp. From the *Transactions of the Royal Society of Literature.* [*see* 241].

1593

Hooker, J. D. On the vegetation of Clarence peak, Fernando Po; with descriptions of the plants collected by Mr. Gustav Mann on the higher parts of that mountain. [London, 1861]. 23 pp. From the *Journal and Proceedings of the Linnean Society,* Vol. 6. [*see* 225].

1594

Hore, Mrs A. B. To Lake Tanganyika in a bath chair. London, Sampson Low, Marston, [etc.], 1886. xi, 217 pp., front. (port.), plate, maps.

1595 a d l

Horton, J. A. B. The medical topography of the West coast of Africa, with sketches of its botany. London, John Churchill, 1859. [iii], 70 pp. [*see* 239].

1596 a d

Houdas, O. V. ed. Le Maroc de 1631 à 1812; extrait de l'ouvrage intitulé Ettordjemân Elmo'arib 'An Douel Elmachriq ou 'Lmaghrib, de Aboulqâsem ben Ahmed Ezziâni; publié et traduit par O. Houdas. Paris, Imprimerie Nationale, Ernest Leroux, 1886. [5], ix, 216, [112] pp. (Publications de l'École des Langues orientales vivantes, Série 2, Vol. 18).

1597 a d

— — Nozhet-elhâdi: histoire de la dynastie saadienne au Maroc (1511–1670), par Mohammed Esseghir ben Elhadj ben Abdallah Eloufrâni; texte arabe publié par O. Houdas. Paris, Ernest Leroux, 1888–[9]. 2 vols. (Publications de l'École des Langues orientales vivantes, Série 3, Vols. 2–3).

1598 a d

Hutchinson, T. J. Impressions of western Africa; with remarks on the diseases of the climate and a report on the peculiarities of trade up the rivers in the Bight of Biafra. London, Longman, Brown, Green, [etc.], 1858. xvi, 313 pp. illus.

a d

Another copy.

1599 a d

— Narrative of the Niger, Tshadda, & Binuë exploration; including a report on the position and prospects of trade up those rivers, with remarks on the malaria and fevers of western Africa. London, Longman, Brown, Green, [etc.], 1855. xi, 267 pp., front. (map).

1600 a d

— Ten years' wanderings among the Ethiopians; with sketches of the manners and customs of the civilized and uncivilized tribes, from Senegal to Gaboon. London, Hurst & Blackett, 1861. xx, 329 pp., front., tables.

1601 a d
Ibn-Khaldun [*i.e.* 'Abd al-Rahman ibn Muhammad, *called* Ibn Khaldun]. Histoire des Berbères et des dynasties musulmanes de l'Afrique septentrionale, par Ibn-Khaldoun; traduite de l'Arabe, par M. Le Baron de Slane. Alger, Imprimerie du Gouvernement, 1852-6. 4 vols.

1602 a d g h j
Italy. Corpo di Stato Maggiore. Possedimenti e protettorati Europei in Africa, 1890, raccolta di notizie geografiche, storiche, politiche e militari sulle regioni costiere africane; seconda edizione. Roma, Voghera Carlo, 1890. x, 196 pp., illus., maps. (L. dal Verme, Rome, 25 June 1890, 2 letters).

1603 d
James, F. L. A journey through the Somali country to the Webbe Shebeyli. [London, 1885]. 10 leaves. Page proofs of the *Proceedings of the Royal Geographical Society,* Vol. 7, pp. 625-41. [*see* p.].

1604 d h i
— The unknown horn of Africa: an exploration from Berbera to the Leopard river; with additions by J. Godfrey Thrupp. London, Liverpool, George Philip, 1888. xiv, 344 pp., front., illus., plates, map (in pocket). (16 Nov. 1888).

1605 a i
Johnson, J. Y. Madeira: its climate and scenery, a handbook for invalid and other visitors, with chapters on the fauna, flora, geology and meteorology; third edition. London, Dulau, 1885. xxxii, 310 pp., plate, maps, tables.

1606 a
Kiepert, H. Erläuterung zu zwei den Fortschritt der afrikanischen Entdeckungen seit dem Alterthum darstellenden Karten. Berlin, Dietrich Reimer, 1873. 16 pp., maps. (Beiträge zur Entdeckungsgeschichte Afrika's, No. 1). Reprinted from *Zeitschrift der Gesellschaft für Erdkunde zu Berlin,* Vol. 8.

1607
Knoblecher, I. Reise auf dem weissen Nil; aus den Original-Manuscripten des General-Vikars von Central-Afrika; bearbeitet von Dr. V. F. Klun. Laibach, Ignaz v. Kleinmayr und Fedor Bamberg, 1851. 47 pp. [*see* 237].

1608 a d
Koelle, S. W. African native literature: or, proverbs, tales, fables, & historical fragments in the Kanuri or Bornu language; to which is added a translation of the above, and a Kanuri-English vocabulary. London, Church Missionary House, 1854. xv, 434 pp.

1609
Koner, W. I. Der Antheil der Deutschen an der Entdeckung und Erforschung Afrika's – II. Erläuterungen zu der die Entdeckungen des 19. Jahrhunderts darstellenden Karte von Afrika, von H. Kiepert. Berlin, Dietrich Reimer, 1874. 59 pp., map. (Beiträge zur Entdeckungsgeschichte Afrika's, 2). Reprinted from *Zeitschrift der Gesellschaft für Erdkunde zu Berlin,* Vol. 8. [*see* 1606].

1610 a d f
Krapf, J. L. Travels, researches, and missionary labours, during an eighteen years' residence in eastern Africa, together with journeys to Jagga, Usambara, Ukambani, Shoa, Abessinia, and Khartum . . . with an appendix and a concise account of geographical researches in eastern Africa . . . by E. G. Ravenstein. London, Trübner, 1860. li, 566 pp., front., (port.), plates, maps, table.

1611
La Motte, – de. Le Nil: première conférence/faite le 16 juillet 1880 à la Société de Géographie de Paris, touchant ses études sur le bassin du Nil [and deuxième conférence (Xbre 1880). Paris, Typ. Tolmer, 1880]. 2 pts. [*see* pq.].

1612
Lander, R. and J. Lander. Journal of an expedition to explore the course and termination of the Niger, with a narrative of a voyage down that river to its termination; second edition. London, Thomas Tegg, 1838. 2 vols., fronts. (port.), illus., plates, map.

1613 a d
Lane, E. W. An account of the manners and customs of the modern Egyptians, written in Egypt during the years 1833, -34, and -35, partly from notes made during a former visit to that country in the years 1825, -26, -27, and -28; the fifth edition, with numerous additions and improvements . . . edited by . . . Edward Stanley Poole. London, John Murray, 1860. xxiii, 619 pp., front., illus.

1614
Lapenna, L. La destitution de Lappena et la réforme judiciaire en Egypte. Genève, 1878. 59 pp. [*see* p.].

1615 a d h i j

Leared, A. A visit to the court of Morocco. London, Sampson Low, Marston, [etc.], 1879. [vii], 86 pp. (Arthur Leared, [London], 3 July 1879].

1616 a d

Lenoir, P. The fayoum: or artists in Egypt; new edition. London, Henry S. King, 1875. vii, 298 pp., front., plates.

1617 i

Leo, Africanus. Ioannis Leonis Africani Africae descriptio IX, Lib. absoluta. Lugd. Batav., apud Elzevir, 1632. 816 pp. (J. Pincherle, Trieste, April 1875).

1618 a d

Lesseps, F. de. Lettres, journal et documents pour servir à l'histoire du Canal de Suez . . . Paris, Librairie Académique Didier, 1875. 2 vols., tables.

1619 g

Lewis, J. A. "Missionaries & missions". Chapter III, from a book partly written on "Sierra Leone and surroundings". Sierra Leone, T. J. Sawyerr, printer, 1885. 47 pp. [*see* p.].

1620 d

Linant de Bellfonds, L. M. A. L'Etbaye, pays habité par les Arabes bicharieh: géographie, ethnologie, mines d'or, par Linant de Bellefonds Bey. Paris, Arthus Bertrand, [1877]. 2 vols., illus., plates. wanting atlas.

1621 a d

— Mémoires sur les principaux travaux d'utilité publique exécutés en Egypte depuis la plus haute antiquité jusqu'à nos jours, par Linand de Bellefonds Bey. Paris, Arthus Bertrand, 1872–3. 2 vols., illus. wanting Vol. 2, and atlas.

1622 a d f h

Livingstone, D. Missionary travels and researches in South Africa, including a sketch of sixteen years' residence in the interior of Africa, and a journey from the Cape of Good Hope to Loanda on the west coast . . . London, John Murray, 1857. x, 711 pp., front., illus., plates, maps.

1623 a d

— and C. Livingstone. Narrative of an expedition to the Zambesi and its tributaries, and of the discovery of the Lakes Shirwa and Nyassa, 1858–1864. London, John Murray, 1865. xv, 608 pp., front., illus., plates, map.

1624 d

Long, C. C. Central Africa, naked truths of naked people: an account of expeditions to the Lake Victoria Nyanza and the Makraka Niam-Niam, west of the Bahr-el-Abiad (White Nile). London, Sampson Low, Marston, [etc.], 1876. xvi, 330 pp., illus., plates, map.

1625 a d h

Lopes de Lima, J. J. Ensaios sobre a statistica das possessões portuguezas na Africa occidental e oriental, na Asia occidental, na China, e na Oceania . . . Lisboa, Imprensa Nacional, 1844–6. Vols. 1–3 (6 pts. in 1), illus., tables, maps. wanting Vols. 4–6.

1626 b d

Macbrair, R. M. The Africans at home: being a popular description of Africa and the Africans condensed from the accounts of African travellers from the time of Mungo Park to the present day. London, Longman, Green, [etc.], 1861. xx, 396 pp., front., illus., plates, map.

1627 h j

Macdonnell, Sir R. G. Our relations with Ashantee: an address delivered before the Royal Colonial Institute, 10th February, 1874. [London, Unwin Brothers, printer, 1874]. 23 pp. (Trelawney Saunders, London, India Office, 7 March 1874). [*see* p.].

1628 a d

M'Leod, J. A voyage to Africa with some account of the manners and customs of the Dahomian people. London, John Murray, 1820. iv, 161 pp., front., plates.

1629 a h

McLeod, L. Travels in eastern Africa, with the narrative of a residence in Mozambique. London, Hurst & Blackett, 1860. 2 vols. fronts., map.

1630 a d f

M'Queen, J. A geographical survey of Africa, its rivers, lakes, mountains, productions, states, population . . . to which is prefixed a letter to Lord John Russell, regarding the slave trade, and the improvement of Africa. London, B. Fellowes, 1840. xxiv, xciv, 303 pp.

1631 d f

Mahmoud-Bey, al-Falaki. Mémoire sur l'antique Alexandrie, ses faubourgs et environs découverts. Copenhague, Imprimerie de Bianco Luno, par F. S. Muhle, 1872. 135 pp. Another copy.

1632 d
Mann, G. Letter from Mr. G. Mann . . . describing his expedition to the Cameroon mountains. London, 1862. 13 pp. From the *Journal and Proceedings of the Linnean Society, Botany,* Vol. 7. [*see* 225].

1633
[Marcel, G.]. Les dernières explorations en Afrique. [Paris, 1875]. [12] pp. From an unidentified journal. [*see* 221].

1634 a d h l
Mariette, A. Notice des principaux monuments exposés dans les galeries provisoires du Musée d'Antiquités égyptiennes de S.A. le Khédive à Boulaq; sixième édition. Le Caire, A. Mourès, 1876. [iii], 31 pp., illus.

1635
[Markham, Sir C. R.]. Geographical results of the Abyssinian expedition. London, [1868]. 38 pp., maps. Reprinted from the *Journal of the Royal Geographical Society,* Vol. 38. [*see* 256].

1636 a d
Maspero, Sir G. C. C. L'archéologie égyptienne. Paris, Maison Quantin, 1887. 318 pp., illus. (Bibliothèque de l'enseignement des beaux-arts).

1637*a* a d h
Massey, G. A book of the beginnings: containing an attempt to recover and reconstitute the lost origines of the myths and mysteries, types and symbols, religion and language, with Egypt for the mouthpiece and Africa as the birthplace. London, Williams & Norgate, 1881. 2 vols., illus.
and
1637*b* d
(i) offprint: A preface to, with extracts from, *A book of the beginnings.* 28 pp. Another copy.
(ii) 3 leaflets.
d
(iii) galley proofs of 'Comparative vocabulary of Sanskrit and Egyptian'. 2 leaves. Another copy.
(iv) manuscript notes. 3 leaves.
(v) autograph letters signed: Gerald Massey, and dated London, 15 April 1881, 27 April [1881?], 1 Oct. 1881, and 10 May 1883.
(vi) holograph by Massey. 7 leaves.

1638 l
— The natural genesis: or, second part of A book of the beginnings . . . London, Williams & Norgate, 1883. 2 vols., illus.

1639 i
Mitchell, L. H. Report on the seizure by the Abyssinians of the geological and mineralogical reconnaissance expedition. Cairo, Printing Office of the General Staff, 1878. xii, 125 pp., map. At head of title: Publications of the Egyptian General Staff. (Cairo, 4 May 1878).

1640
— and C. I. Graves. Reconnaissance des anciennes mines de Hammamat – Le pays des Somalis Mijjertains. Le Caire, Secrétariat de la Société Khédiviale de Géographie, 1879. 40 pp. From the *Bulletin de la Société Khédiviale de Géographie,* Vol. 6. [*see* p.].

1641 a d
Mourão-Pitta, C. A. Du climat de Madère et de son influence thérapeutique . . . Montpellier, Typographie de Boehm, 1859. [i], 262 pp. [*see* 240].

1642 a d
Muhammad ibn 'Umar, al-Tūnusī. Voyage au Darfour, par le Cheykh Mohammed ebn-Omar el-Tounsy; traduit de l'Arabe par le Dr Perron; publié par les soins de M. Jomard . . . Paris, Benjamin Duprat, 1845. lxxxviii, 492 pp., front. (port.), plates, map.

1643 a d
Murray, E. Sixteen years of an artist's life in Morocco, Spain, and the Canary islands. London, Hurst & Blackett, 1859. 2 vols., fronts.

1644 a d
Murray, H. Discovery and adventure in Africa, with a narrative of recent exploring expeditions. London, T. Nelson, 1857. 496 pp., front., illus., plates, maps.

1645 h
Nachtigal, G. and Mohammed Moktar. Voyage au Wadaï – Notes sur le pays de Harrar. Le Caire, Secrétariat de la Société Khédiviale de Géographie, 1877. 305–412 pp., map. From the *Bulletin trimestriel de la Société Khédiviale de Géographie,* Series 1, Vol. 1/4, No. 4. [*see* p.].
Another copy.
1646
Ninet, J. Dégénérescence du coton égyptien "mako-jumel". Paris, Librairie Fischbacher, [1887]. 77 pp., front. [*see* p.].
1647 e
Norden, F. L. Travels in Egypt and Nubia; translated from the original . . . and enlarged with observations from ancient and modern authors, by Peter Templeman. London, Lockyer Davis & Charles Reymers, 1757. 2 vols. (in 1), front., plates.
1648 a b d
Norton, Mrs C. E. S. ed. A residence at Sierra Leone, described from a journal kept on the spot, and from letters written to friends at home, by a Lady. London, John Murray, 1849. xii, 335 pp., illus.
1649 d
Owen, W. F. W. Narrative of voyages to explore the shores of Africa, Arabia, and Madagascar, performed in H.M. ships Leven and Barracouta. London, Richard Bentley, 1833. 2 vols., front., illus., plates, maps.
1650
Park, M. Travels in the interior of Africa. Edinburgh, Adam & Charles Black, 1860. xvi, 380 pp., front., plates.
1651 d
Paton, A. A. A history of the Egyptian revolution, from the period of the Mamelukes to the death of Mohammed Ali . . . second edition, enlarged. London, Trübner, 1870. 2 vols.
1652 a i
Paulitschke, P. Die geographische Erforschung der Adâl-Länder und Harâr's in Ost-Afrika. Leipzig, Paul Frohberg, 1884. vii, 109 pp. (Vienne, 19 août 1884).
1653 h i l
— Major Heaths und Leutnant Peytons Reise von Hárár nach Bérbera, Juni 1885. [Gotha], 1886. 65–7 pp., map. Reprinted from *Dr. A. Petermanns Mitteilungen,* 1886, No. 3. (Vienna, 18th April 1886). [*see* p.].
1654 a d h i
Peacock, G. The Guinea: or Gold Coast of Africa. Exeter, William Pollard, 1880. 52 pp., front. (map), map. [*see* p.].
1655
Penning, W. H. Notes on the gold fields of the Transvaal. London, South African Syndicate Company, 1883. 19 pp. Reprinted from *De Volksstem,* 27 Jan. 1883. [*see* p.].
1656
Perpetua, J. Géographie de la régence de Tunis. Tunis, Vittorio Finzi, 1883. 47 pp. [*see* p.].
1657
Perry, H. A. The future of justice in Egypt. London, P. S. King, 1881. [i], 38 pp. [*see* p.].
1658 a d
Petherick, J. Egypt, the Soudan and central Africa, with explorations from Khartoum on the White Nile to the regions of the Equator: being sketches from sixteen years' travel. Edinburgh, London, William Blackwood, 1861. xii, 482 pp., illus., map.
1659 a d
— and Mrs B. H. Petherick. Travels in central Africa, and explorations of the western Nile tributaries. London, Tinsley, 1869. 2 vols., fronts., (port.), illus., plates, map.
1660 a e
Pietri, J. Les voyages de M. Giraud aux lacs de l'Afrique centrale. Le Caire, Imprimerie Nationale, 1888. 697–705 pp. From the *Bulletin de la Société Khédiviale de Géographie,* Series 2, Vol. 12, Supplement. [*see* p.].
1661
Playfair, Sir R. L. A bibliography of Algeria. London, John Murray, 1888. [v], 129–430 pp. (Royal Geographical Society supplementary papers, Vol. 2, No. 2).
1662 a d h
— Handbook for travellers in Algeria and Tunis, Algiers, Oran, Tlemçen, Bougie,

Constantine, Tebessa, Biskra, Tunis, Carthage, etc.; fourth edition. London, John Murray, 1890. xiii, 360 pp.

1663

Purdy, E. S. Psychrometrical observations taken at Fascher, Darfour . . . expedition of reconnaissance 1876. Cairo, Printing Office of the General Staff, 1877. 15 pp. At head of title: Publication of the Egyptian General Staff. [*see* p.].

1664 a d h

Reade, W. W. Savage Africa: being the narrative of a tour in equatorial, south-western, and north-western Africa, with notes on the habits of the gorilla, on the existence of unicorns and tailed men, on the slave-trade . . . London, Smith, Elder, 1863. xv, 587 pp., front., plates, map.

1665 d

Ricketts, H. I. Narrative of the Ashantee war, with a view of the present state of the Colony of Sierra Leone. London, Simpkin & Marshall, 1831. [ii], 221 pp., front., plates. [*see* 238].

1666 a d

Ritter, K. Géographie générale comparée, ou étude de la terre dans ses rapports avec la nature et avec l'histoire de l'homme . . . traduit de l'allemand, par E. Buret et Édouard Desor. Bruxelles, Etablissement encyclographique, 1837. [i], 620 pp.

1667 a d

Robecchi Brichetti, G. Une excursion à Siwa. Le Caire, Imprimerie Nationale, 1889. 83–118 pp. From the *Bulletin de la Société Khédiviale de Géographie,* Series 3, Vol. 2. Title on wrapper: Excursion à l'oasis de Siwa. [*see* p.].

1668 d i

Schweinfurth, G. Die ältesten Klöster der Christenheit (St. Antonius und St. Paulus). [n.p., 1877]. 275–316 pp., illus. (Cairo, October 1877). [*see* p.].

1669

— La terra incognita dell'egitto propriamente detto. Milano, Lombarda, 1878. 48 pp., illus., map. Reprinted from *L'Esploratore,* Vol. 2, Nos. 4–6. [*see* p.].

1670 a d h

Serpa Pinto, A. A. de R. How I crossed Africa, from the Atlantic to the Indian ocean, through unknown countries, discovery of the great Zambesi affluents, &c.; translated from the author's manuscript by Alfred Elwes. London, Sampson, Low, Marston, [etc.], 1881. 2 vols., front., illus., plates, maps.

1671 d l

Shepherd, A. F. The campaign in Abyssinia. Bombay, 'Times of India' Office, 1868. xxvi, 388, 1 pp., front. (map).

1672

Silver, S. W., & Co. Handbook to South Africa, including the Cape Colony, Natal, the diamond fields, the Transvaal, Orange Free State, etc., also a gazetteer; second edition. London, S. W. Silver, Office of 'The Colonies and India', 1879. xi, 567 pp., front. (map), plates, tables.

1673 a d i

Smith, J. L. C. Narrative of the discovery of the great central lakes of Africa: Tanganyika, Victoria Nyanza, Albert Nyanza, and Nyassa. Halifax, F. King. 1877. 32 pp., front. (map).

1674 a d

Smith, W. A new voyage to Guinea, describing the customs, manners, soil, climate, habits, buildings . . . London, John Nourse, [1744]. iv, 276, [8] pp., plates.

1675

Smyth, C. P. Madeira meteorologic: being a paper . . . read before the Royal Society, Edinburgh, on the 1st May 1882. Edinburgh, David Douglas, 1882. viii, 84 pp., front., illus., plate.

1676

Sociedade de Geographia de Lisboa. Ao povo Portuguez em nome da honra, do direito, do interesse e do futuro da patria a Commissão do Fundo africano. Lisboa, Imprensa Nacional, 1881. 19 pp., map. [*see* pq.].

1677 a d

— La question du Zaire: droits du Portugal, memorandum; édition française. Lisbonne, Lallemant, 1883. 80 pp. [*see* p.].

1678 d

Speke, J. H. Captain J. H. Speke's discovery of the Victoria Nyanza lake, the supposed source of the Nile; from his Journal. Edinburgh, 1859. 339–57, 391–419, 565–82 pp., map. From *Blackwood's Edinburgh magazine,* Sept.–Nov. 1859, Vol. 86. Part 1

entitled: Journal of a cruise on the Tanganyika lake, central Africa. [*see* 9].
d
Another copy of Part 3. [*see* 238].
1679 a d h
— Journal of the discovery of the source of the Nile. Edinburgh, London, William Blackwood, 1863. xxxi, 658 pp., front. (port.), illus., plates, maps.
1680 a d
— What led to the discovery of the source of the Nile. Edinburgh, London, William Blackwood, 1864. x, 372 pp., front., maps.
1681 d h l
Stanley, Sir H. M. The Congo and the founding of its Free State: a story of work and exploration. London, Sampson Low, Marston, [etc.], 1885. 2 vols., fronts., illus., plates, maps.
1682 a d h
— How I found Livingstone: travels, adventures, and discoveries in central Africa, including four months' residence with Dr. Livingstone; second edition. London, Sampson Low, Marston, [etc.], 1872. xxiii, 736 pp., front. (port.), illus., maps.
1683
— Speech of Mr. H. M. Stanley at the Stanley Club in Paris, October 19th 1882. London, J. Miles, 1882. 24 pp. [*see* p.].
1684 a d
— Through the dark continent: or the sources of the Nile around the great lakes of equatorial Africa, and down the Livingstone river to the Atlantic ocean. London, Sampson Low, Marston, [etc.], 1878. 2 vols., fronts. (ports.), illus., plates, maps.
1685 a d
Taylor, E. M. Madeira: its scenery, and how to see it, with letters of a year's residence, and lists of the trees, flowers, ferns, and seaweeds. London, Edward Stanford, 1882. xvi, 261 pp., front., map.
1686 a d
Travassos Valdez, F. Six years of a traveller's life in western Africa. London, Hurst & Blackett, 1861. 2 vols., front., illus., plates.
1687 d
Tristram, H. B. The great Sahara: wanderings south of the Atlas mountains. London, John Murray, 1860. xv, 435 pp., front., illus., plates, maps.
1688 a d
Tucker, S. Abbeokuta: or, sunrise within the tropics: an outline of the origin and progress of the Yoruba Mission; sixth edition. London, James Nisbet, 1858. vii, 278 pp., front., plates, map.
1689 a d f h
Tuckey, J. H. Narrative of an expedition to explore the river Zaire, usually called the Congo, in South Africa, in 1816 . . . to which is added the Journal of Professor Smith, some general observations on the country and its inhabitants . . . London, John Murray, 1818. [7], lxxxii, 498 pp., front. (map), illus., plates.
1690 a d i
Vaux, W. S. W. On the knowledge the ancients possessed of the sources of the Nile. [London], 1864. 32 pp., front. (map). Reprinted from the *Transactions of the Royal Society of Literature,* N.S. Vol. 8. (July 1864). [*see* p.].
Another copy. [*see* 241].
1691 a
— On recent researches at Budrum (the ancient Halicarnassus), Branchidae, and Cnidus, by C. T. Newton. [London, 1859]. 55 pp. Reprinted from the *Transactions of the Royal Society of Literature,* Series 2, Vol. 6. [*see* 241].
1692 i j
Verme, L. dal. Il paese dei Somali: memoria. Roma, Tipografia delle Mantellate, 1889. 55 pp., map, tables. (L. dal Verme, Rome, 3 July 1889). [*see* p.].
1693
Verneuil, V. Mes aventures au Sénégal: souvenirs de voyage. Paris, Librairie Nouvelle, Jaccottet, Bourdilliat, 1858. [iii], 282 pp.
1694 i
Wace, H. T. Palm leaves from the Nile: being a portion of the diary of a wanderer in Egypt; printed for private circulation only. Shewsbury, Leake & Evans, printer, 1865. 84 pp.
d h k
Another copy.

h i
Third copy.

1695
Wahed. Lettre au Très-Honorable Georges G. Goschen, délégué officieux de quelques créanciers du dualisme Khédivial auprès de S. A. Ismaël Pacha Vali de la Province d'Egypte. La Haye, Typographie Cosmopolite et Internationale, 1877. 80 pp. [*see* p.].

1696 j
Wall, T. A. Colonial administration on the west coast of Africa: an appeal for justice addressed to the peers and Commons [of] Great Britain and Ireland. London, Perkins Bacon, printer, 1885. [i], 35 pp. (T. A. Wall, 28 July 1885). [*see* p.].

1697 a d
Wallace, Sir D. Mack. Egypt and the Egyptian question. London, Macmillan, 1883. x, 521 pp.

1698
Ward, H. My life with Stanley's rear guard; with a map by F. S. Weller. London, Chatto & Windus, 1891. viii, 163 pp.

1699
Wesleyan Missions. Report of the Wesleyan Missions in the Gold Coast, Yoruba, & Popo district . . . for the year 1881. Faversham, R. Lancefield, printer, 1882. 47 pp. [*see* p.].

1700
Whately, M. L. Among the huts in Egypt: scenes from real life; third edition. London, Seeley, Jackson & Halliday, 1873. viii, 344 pp., front., illus.

1701
— Scenes from life in Cairo: a glimpse behind the curtain. London, Seeley, Jackson & Halliday, 1883. viii, 293 pp., front.

1702 a d h
White, R. and J. Y. Johnson. Madeira, its climate and scenery: a handbook for visitors; second edition. Edinburgh, Adam & Charles Black, 1860. xv, 338 pp., front., plates, map, tables.

1703
Whitehouse, F. C. The latest researches in the Moeris basin. [London], 1884. 6 pp., map. Reprinted from the *Proceedings of the Royal Society and Monthly record of geography,* Oct. 1884. [*see* p.].

1704
— Moeris: the wonder of the world. New York, John Wiley, 1885. 16 pp., illus., maps. Reprinted from *School of mines quarterly,* Nov. 1884. [*see* p.].

1705
Wild, G. Von Kairo nach Massaua: eine Erinnerung an Werner Munzinger. Olten, Buchdruckerei des "Volksblatt vom Jura", 1879. xxxii, 79 pp., front. (port.), plates, map, tables. [*see* p.].

1706 d k
Wilkinson, Sir J. G. A handbook for travellers in Egypt, including descriptions of the course of the Nile to the second cataract, Alexandria, Cairo, the pyramids, and Thebes, the overland transit to India, the peninsula of Mount Sinai, the oases, &c. . . . a new edition, with corrections and additions. London, John Murray; Paris, Galignani, Stassin & Xavier; Malta, Muir, 1858. xxiv, 439 pp., illus., map.

1707 a d
Wilson, J. L. Western Africa: its history, condition, and prospects. New York, Harper & Brothers, 1856. 527 pp., front., illus., map.

1708 a d
Winterbottom, T. An account of the native Africans in the neighbourhood of Sierra Leone; to which is added, an account of the present state of medicine among them. London, C. Whittingham, printer, 1803. 2 vols., front., plates, tables.

1709 a d h j
Wylde, A. B. '83 to '87 in the Soudan, with an account of Sir William Hewett's mission to King John of Abyssinia. London, Remington, 1888. 2 vols., front. (map). (A. B. Wylde, Grove Park, Chiswick, 1 Oct. 1888).

1710 a d
Zotenberg, H. T. La chronique de Jean, Évêque de Nikiou: notice et extraits. Paris, Imprimerie Nationale, 1879. [iii], 264 pp. Reprinted from *Journal asiatique,* 1877, Series 7, Vols. 10, 12–13.

1711
Zweifel, J. and M. Moustier. Voyage aux sources du Niger. Marseille, Typ. et Lith.

Barlatier-Feissat, 1880. [v], 164 pp., plates, map, tables. At head of title: Expédition C. A. Verminck.

1712-60 North America

1712 d
Aubertin, J. J. A fight with distances: The States, the Hawaiian islands, Canada, British Columbia, Cuba, the Bahamas. London, Kegan Paul, Trench, 1888. ix, 352 pp., front., plates, maps.

1713 i
Batcheler, H. P. Jonathan at home: or, a stray shot at the Yankees. London, W. H. Collingridge, 1864. viii, 287 pp. (11 April 1865).

1714 i
Blake, W. P. Geographical notes upon Russian America and the Stickeen river . . . Washington, Government Printing Office, 1868. 19 pp., illus., map. (Commodore C. R. R. Rodgers, Jan. 1873). [*see* p.].

1715 d
[Bodman, A. H. intro. Bodman's hand-book of Chicago. Chicago, 1859]. viii, 131 pp. [*see* 244].

1716
Bohn's hand-book of Washington. Washington, Casimir Bohn, 1860. 133 pp., front., map. [*see* 247].

1717
The Canadian Pacific railway. Montreal, 1887. 48 pp., front., illus. [Rubber stamped: . . . Tacoma, W.T.]. [*see* p.].

1718 d i
Carega di Muricce, F. Studi di economia rurale americana applicati all'Italia; 2a edizione. Rocca S. Casciano, Stab. Tipografico di F. Cappelli, 1874. 15 pp., illus. [*see* p.].

1719
Chandless, W. A visit to Salt Lake: being a journey across the plains and a residence in the Mormon settlements at Utah. London, Smith, Elder, 1857. xii, 346 pp., front. (map).

1720
Chapman, S. Hand book of Wisconsin; second edition. Milwaukee, S. Chapman, 1855. 117 pp. [*see* 247].

1721 a
Clarke, H. Researches in prehistoric and protohistoric comparative philology, mythology, and archaeology in connection with the origin of culture in America and the Accad or Sumerian families. London, Trübner, 1875. xi, 74 pp. [*see* p.].

1722 e
Degroot, H. Sketches of the Washoe silver mines . . . soil, climate and mineral resources of the country east of the Sierra. San Francisco, Hutchings & Rosenfield, 1860. 24 pp. [*see* 246].

1723
Dixon, W. H. New America; sixth edition. London, Hurst & Blackett, 1867. 2 vols., fronts., plates.

1724 d
Goodrich, C. A. A history of the United States of America. Boston, Hickling, Swan & Brewer, 1859. 440 pp., front., maps, illus.

1725 d l
Gunnison, J. W. The Mormons: or, Latter-Day Saints, in the valley of the Great Salt lake. Philadelphia, Lippincott, Grambo, 1852. 168 pp., front., illus.

1726 i
Hatton, J. Henry Irving's impression of America. London, Sampson Low, Marston, [etc.], 1884. 2 vols. (Henry Irving, 3 June 1884).
1727 a d h
Haven, S. F. Archaeology of the United States. [Washington, Smithsonian Institution, 1856]. [iv], 168 pp. (Smithsonian contributions to knowledge). [*see* q.].

1728 a l
Hayden, F. V. Contributions to the ethnography and philology of the Indian tribes of the Missouri valley. Philadelphia, C. Sherman, printer, 1862. [iv], 231-461 pp., plates, map. Reprinted from the *Transactions of the American Philosophical Society,* N.S. Vol. 12, No. 2.
1729 d
Horn, H. B. Horn's overland guide, from the U.S. Indian Sub-Agency, Council Buffs . . . to the city of Sacramento in California. New York, J. H. Colton, 1853. 78 pp., map.
1730 d f
Hunter, W. S., jr. Hunter's paroramic guide from Niagara Falls to Quebec. Boston, John P. Jewett; Cleveland, Ohio, Henry P. B. Jewett, 1857. 67 pp., illus., map.
1731 b
[Mackay, C.]. The Mormons: or, Latter-Day Saints . . . third edition. London, [Vizetelly, printer], 1852. 320 pp., front., illus.
1732 d h
Midgley, R. L. [*pseud. i.e.* David Pulsifer]. Sights in Boston and suburbs: or, guide to the stranger. Boston, John P. Jewett, 1856. [xi], 225 pp., front., illus., maps.
1733 d
Montreal. [City of Montreal guide. Montreal, *c.* 1860]. 36 pp., illus., plates. [*see* 244].
1734
Mowry, S. The geography and resources of Arizona & Sonora: an address before the American Geographical & Statistical Society. Washington, Henry Polkinhorn, printer, 1859. 48 pp. wanting pp. 45–6. [*see* 246].
1735 i
— The geography and resources of Arizona and Sonora . . . new edition. San Francisco, New York, A. Roman, 1863. 124 pp., front. (map).
1736 d i
— Arizona and Sonora: the geography, history, and resources of the Silver region of North America . . . third edition. New York, Harper, 1864. 251 pp., front.
1737 d
Niagara. The falls of Niagara: being a complete guide . . . London, [etc.], T. Nelson, 1859. 64 pp. [*see* 247].
1738
North American railway & steamboat guide. Baltimore, Sherwood & White, [*c.* 1860]. 202 pp. [*see* 247].
1739
Nil.
1740
Pike, Z. M. Exploratory travels through the western territories of North America . . . performed in the years 1805, 1807 . . . London, Longman, Hurst, [etc.], 1811. xx, 436 pp., front. (map), map, tables.
1741 d
Pratt, O. Absurdities of immaterialism: or, a reply to T. W. P. Taylder's pamphlet, entitled, 'The materialism of the Mormons' . . . examined and exposed. [Liverpool, R. James, printer, 1849]. 32 pp. [*see* 243].
1742 d
Pratt, P. P. Marriage and morals in Utah: an address. Liverpool, Orson Pratt, [etc.], 1856. 8 pp. [*see* 243].
1743 d
Quebec. The Quebec guide . . . places of interest in and about the city and country adjacent . . . Quebec, P. Sinclair, bookseller, 1857. 48 pp., front. (table). wanting pp. 27–8. [*see* 244].
1744
Randolph, E. Address on the history of California from the discovery of the country to the year 1849 . . . tenth anniversary of the admission of the State of California into the Union. San Francisco, Alta California Job Office, printer, 1860. 72 pp., maps. [*see* 246].

1745 a d

Remy, J. Voyage au pays des Mormons: relation, géographie, histoire naturelle . . . moeurs et coutumes. Paris, E. Dentu, 1860. 2 vols. (in 1), fronts., plates, map.

1746 d f

Richards, T. A. Appleton's Illustrated hand-book of American travel: a full and reliable guide by railway, steamboat, and stage. New York, D. Appleton; London, Trübner, 1860. 2 vols. (in 1); fronts. (maps), illus., maps.

1747 d

Russell, W. S. Pilgrims memorials, and guide to Plymouth . . . second edition. Boston, Crosby, Nichols, 1855. [ii], 203 pp., front. (map), plates.

1748 d

Seward, W. H. The state of the country: speech . . . in the United States Senate, February 29, 1860. [Washington, D.C., 1860]. 27 pp. [*see* 246].

1749 b i

Soulé, F., J. H. Gihon and J. Nisbet. The annals of San Francisco; containing a summary of the history . . . and present condition of California. New York, D. Appleton, 1855. [ii], 824 pp., front., illus., plates, map. (Donald Davidson, San Francisco).

1750

Table rock album and sketches of the Falls and scenery adjacent. Buffalo, Steam Press of E. R. Jewett, 1859. 120 pp., illus. [*see* 244].

1751

Taylor, B. Eldorado: or, adventures in the path of empire: comprising a voyage to California, via Panama . . . eighteenth edition. New York, G. P. Putnam, 1859. xiv, 444 pp., front.

1752 d

Thomas, G. F., ed. Appleton's Illustrated and steam navigation guide . . . throughout the United States and the Canadas. New York, D. Appleton, 1860. 264 pp., maps, tables.

1753 d h

Tocqueville, A. de. De la démocratie en Amérique . . . treizième édition. Paris, Pagnerre, 1850. 2 vols. (in 1).

1754 d f

The Traveller's steamboat and railroad guide to the Hudson river . . . New York, Phelps & Watson, [1857]. 50 pp., illus. [*see* 245].

1755 i

United States of America. General Land Office. Instructions to the surveyors general of the public lands of the United States, for those surveying districts established in and since the year 1850. Washington, A. O. P. Nicholson, printer. viii, 35, 56 pp., map, tables. (S. C. Stambaugh, Surveyor Genl. U.S., Salt Lake City, 18 Sept. 1860).

1756 e

— U.S. Army. Engineering Department. Report of the geological exploration of the fortieth parallel. Volume 3: Mining industry by James D. Hague, with geological contributions by Clarence King . . . Washington, Government Printing Office, 1870. xv, 647 pp., front., plates.

1757

Utah, Territory of. Legislative Assembly. Acts, resolutions and memorials, passed at the several annual sessions . . . Great Salt Lake City, Joseph Cain, public printer, 1855. 460 pp. [imperfect].

1758

— — Acts, resolutions and memorials passed . . . during the ninth annual session, for the years 1859–60. Great Salt Lake City, John S. Davis, printer, 1860. iv, 44 pp. [*see* 244].

1759 d

Wineberger, J. A. The tomb of Washington at Mount Vernon . . . Washington, Thomas McGill, 1858. 70 pp., front. (port.), plates. [*see* 246].

1760 d

Wright, C. W. A guide manual to the mammoth cave of Kentucky. Louisville, Ky, Bradley & Gilbert, printers, 1860. 61 pp. [*see* 244].

1761-905 Middle and South America, West Indies

1761
Anonymous. Brazil (From our special correspondent at Rio Janeiro. II). London, 1880. 1 leaf. From *The Times,* 20 Sept. 1880, p. 4. [*see* p.].

1762
— The Paulo Affonso falls, San Francisco river, Brazil. London, 1881. 1 leaf. From *The Graphic,* 7 May 1881, p. 448. [*see* p.].

1763 a d
Abreu e Lima, J. I. Compendio da historia do Brasil . . . edição em um volume. Rio de Janeiro, Eduardo y Henrique Laemmert, [1852]. vii, 352 pp.

1764 a d
Agassiz, L. and Mrs E. C. Agassiz. A journey in Brazil. London, Trübner, [1868]. xix, 540 pp., front., plates.

1765 a d
Alaux, G. d'. L'empereur Soulouque et son empire. Paris, Michel Lévy, 1856. [iv], 286 pp.

1766 d
Alincourt, L. d'. Memoria sobre a viagem do porto de Santos á cidade de Cuyabá . . . Rio de Janeiro, Typographia Imperial e Nacional, 1830. xii, 198 pp. [*see* 251].

1767 a d
Amunátegui, M. L. Descubrimiento i conquista de Chile. Santiago de Chile, Imprenta Chilena, 1862. ix, 532 pp.

1768 d
Andrada e Silva, J. B. de, and M. F. Ribeiro de Andrada. Viagem mineralogica na Provincia de S. Paulo. [Rio de Janeiro, 186-?]. 56 pp., *and* Indice de legislação portugueza as minas, 16 pp. [*see* 242].

1769 a d l
Andrews, J. Journey from Buenos Ayres, through the provinces of Cordova, Tucuman, and Salta, to Potosi . . . in the years 1825-26. London, John Murray, 1827. 2 vols.

1770 d f g
Assis Martins, A. de, and J. Marques de Oliveira. Almanak administrativo, civil e industrial da provincia de Minas-Geraes para o anno de 1864. Rio de Janeiro, Typographia da Actualidade, 1864. [vii], 435 pp., table.

1771 i
Aubertin, J. J. Carta dirigida aos srs. habitantes da provincia de S. Paulo. S. Paulo, Typographia Litteraria, 1862. [iii], 34 pp. [*see* 242].

1772 d
— Eleven days' journey in the province of Sao Paulo . . . letter addressed to His Excellency the Baron of Piracicaba. London, [Effingham Wilson, printer], 1866. 40 pp. [*see* 251].

1773 h i
— A flight to Mexico. London, Kegan Paul, Trench, 1882. [vii], 325 pp.

1774 k
— By order of the sun to Chile to see his total eclipse April 16, 1893. London, Kegan Paul, Trench, Trübner, 1894. [xii], 152 pp., front., plates.

1775
Autenrieth, E. L. A topographical map of the isthmus of Panama [and] Catalogue of maps, charts, books, etc., published by J. H. Colton. New York, J. H. Colton, 1851. 14, [4], 16 pp., maps. [*see* p.].

1776

Azevedo, D. de. Ligeiro esboço da viagem de inauguração ao Rio S. Francisco. Bahia, Typ. de Camillo de Lellis Masson, 1867. 29 pp. [*see* 252].

1777 a d

Azevedo, M. de. Pequeno panorama ou descripção dos principaes edificios da cidade do Rio de Janeiro. Rio de Janeiro, Typographia de F. de Paulo Brito, 1861–2. 3 vols. (in 1).

1778 a d

Barros Arana, D. Compendio de historia de América. Santiago, Imp. del Ferrocarril, 1865. 2 vols.

1779 a d h

Bates, H. W. The naturalist on the river Amazons: a record of adventures, . . . sketches of Brazilian and Indian life. London, John Murray, 1863. 2 vols., fronts., illus., plates, map.

1780

Bates, Hendy & Co. The river Plate (South America) as a field for emigration . . . London, Bates, Hendy, [etc.], [186–]. 35 pp. [*see* 255].

1781 a d

Bernard, L. B. Le rio Parana: années de séjour dans la République Argentine. Paris, Grassart, [etc.], 1864. [iv], 295 pp.

1782 a d i

Bollaert, W. Antiquarian, ethnological and other researches in New Granada, Equador, Peru and Chile . . . London, Trübner, 1860. [ii], 279 pp., front., plates. (R. G. Latham, 19 Aug. 1860).

1783

Bomfin Espindola, T. do. Geographia physica, politica, historica e administrativa da Provincia das Alagoas. Maceió, Typographia do Jornal de Maceió, 1860. 33 pp., tables. [*see* 250].

1784 a d l

Bonnycastle, R. H. Spanish America: or a descriptive, historical, and geographical account of the dominions of Spain in the western hemisphere, continental & insular. Philadelphia, Abraham Small, 1819. 482 pp., front. (map).

1785 a d i

Bossi, B. Viage pintoresco por los rios Paraná, Paraguay, Sn Lorenzo, Cuyabá . . . con la descripcion de la provincia de Mato Grosso . . . Paris, Dupray de la Mahérie, 1863. xi, 155 pp., front. (port.), illus., plates, map. (21 Nov. 1868).

1786

Brazil. Commando em chefe de todas as forças brazileiras em operáçoes contra o governo do Paraguay: assumpçáo 14 de janeiro de 1869; ordem do dia N. 272. [Rio de Janeiro], Typ. do Exercito, Imp. L. C. O. Guimaraes, 1869. 15 pp. [*see* 250].

1787 d i

Butterfield, C. United States and Mexican mail steamship line, and Statistics of Mexico. New York, J. A. H. Hasbrouck, 1860. 109, [ii], 159 pp., maps. (H. S. Bruton).

1788 a d l

[Caldreon de la Bara, Mme]. Life in Mexico, during a residence of two years in that country . . . with a preface by W. H. Prescott. London, Chapman and Hall, 1843. xii, 436 pp.

1789 a d

Capanema, G. S. de. Decomposição dos penedos no Brasil. Rio de Janeiro, Typographia Perseverança, 1866. 32 pp. [*see* 249].

1790 d i

Carega di Muricce, F. Storia ed ascensione del Popocatepetl 1871: due lettere indirizzate al Comm. Prof. Giuseppe Meneghini. Firenze, Tipografia della Gazzetta d'Italia, 1874. 12 pp. Reprinted from *Gazzetta d'Italia*. [*see* p.].

1790.1 l

— Studi di economia rurale americana, applicati all'Italia; 2a. edizione. Rocca S. Casciano, Stab. Tipografico di F. Cappelli, 1874. 15 pp. (Memoria letta nel Marzo 1873 all'Accademia dei Georgofili). [*see* p.].

1791

[Central Argentine Railway Company]. Letters concerning the country of the Argentine Republic (South America) being suitable for emigrants and capitalists to settle in . . . second issue. London, Waterlow, printers, 1869. [ii], 34 pp. [*see* 251].

1792 a d

Charlevoix, P. F. X. de. Histoire du Paraguay. Paris, Desaint & Saillant, [etc.], 1756. 3 vols., illus., maps.

1793 l

Coghlan, J. Waterworks, drainage, and street improvements: report. Buenos Ayres, Standard Printing Office, 1869. 54 pp., tables, maps. At head of title: City improvements in Buenos Ayres. wanting pp. 41–6. [*see* 253].

1794 l

Companhia União e Industria. Relatorio apresentado a assemblea geral dos accionistas da Companhia União e Industria em 3 de fevereiro de 1868. Rio de Janeiro, Typ. do Correio Mercantil, 1868. 17 pp., tables; *and* Annexos, 18 pp. [*see* 249].

1795 d

Correa do Couto, A. Dissertação sobre o actual governo da Republica do Paraguay. Rio de Janeiro, Typographia do Imperial Instituto artistico, 1865. 104 pp., plates. [*see* 242].

1796 e

Corwine, A. B. The Panama massacre: a collection of the principal evidence and other documents, including the report of Amos B. Corwine, Esq., U.S. Commissioner. Panama, New Granada, Office of the Star and Herald, printer, 1857. [ii], 69 pp. [*see* 246].

1797 a d

Costa Aguiar, A. A. da. O Brazil e os Brazileiros. Santos, Typographia Commercial, 1862. [ii], 150 pp. [*see* 251].

1798

Costa Batinga, M. J. da. Mappa da exportação pela mesa de rendas provinciaes da muita leal e valerosa cidade do Penedo do Rio de São Francisco, provincia das Alagôas. Penedo, 1853. 1 sheet. [*see* p.].

1799 a d i

Costa Honorato, M. da. Diccionario topographico, estatistico e historico da provincia de Pernambuco. Recife, Typographia Universal, 1863. [viii], 188 pp. [*see* 255].

1800

Cox, G. E. Viaje a las rejiones septentrionales de la Patagonia. Santiago, 1863. [100 pp.]. From *Anales, Universidad de Chile,* Vol. 23, pp. 3–103. [*see* 248].

1801 d

Cuadra, P. L. Apuntes sobre la jeografía física i política de Chile. Santiago, Imprenta Nacional, 1868. 2 parts (in 1).

1802 d

Dabadie, F. A travers . . . l'Amérique du sud . . . deuxième édition. Paris, Ferdinand Sartorius, 1859. [iii], 7, 387 pp.

1803 a d

— Récits et types américains. Paris, F. Satorius, 1860. [iii], 384 pp.

1804 a d i

Denis, F. Brazil, por Fernando Denis – Colombia e Guyanas, por M. C. Famin; tradusida do frances. Vol. 1. Lisboa, Typ. de L. C. da Cunha, 1844. Half-title: Descripção historica do Brazil. (Maria Gabriela).

1805 h

Dent, H. C. A year in Brazil with notes on the abolition of slavery, the finances of the empire, religion . . . London, Kegan Paul, Trench, 1886. xix, 444 pp., front., plates, maps.

1806 i

Domeyko, I. Elementos de mineralojia . . . en particular de las de Chile; segunda edicion. Santiago, Imprenta del Ferrocarril, 1860. xi, 432 pp. (Santiago, 1869). [*see* 248].

1807

— Mémoire sur le terrain tertiaire et les lignes d'ancien niveau de l'océan du sud, aux environs de Coquimbo (Chili). [Paris, Imprimé par E. Thunot, 1848]. 10 pp., map. From *Annales des Mines,* Série 4, Vol. 14. [*see* 507].

1808

— Mineralojia de Chile; segundo apéndice a la 2a edicion de La mineralojia de Don Ignacio Domeyko. Santiago de Chile, Imprenta Nacional, 1867. 52 pp. [*see* 250].

1809 a d h i

[Duffy, J. W.]. A hand book to Valparaiso, containing the laws and regulations of the port . . . Valparaiso, W. Helfmann's "Universo" Printing Office, 1862. viii, 64 pp., front., plate, map. (H. W. Rouse).

1810 a d

Dunlop, C. Brazil as a field for emigration: its geography, climate, agricultural capabilities, and the facilities afforded for permanent settlement. London, Bates, Hendy, [etc.]. 1866. 32 pp. [*see* 251].

1811

Erasmo, pseud. Ao imperador: cartas. Rio de Janeiro, [Typ. de Mello], 1865. viii, 84 pp. [*see* 253].

1812 a d

Ewbank, T. Life in Brazil: or, the land of the cocoa and the palm. London, Sampson Low, 1856. 469 pp., illus.

1813 a d

Expilly, C. Les femmes et les moeurs du Brésil. Paris, Charlieu et Huillery, 1863. xii, 450 pp.

1814 a d

— Le Brésil tel qu'il est; troisième édition. Paris, Charlieu et Huillery, 1864. 383 pp.

1815

Exposição bahiana. Catalogo dos objectos apresentados . . . em 1866. Bahia, Typographia de Tourinho, 1866. [i], 52 pp. [*see* 255].

1816

[Ferreira de Almeida, J. A.]. Impressões de um filho do provo escriptas ao correr de penna no dia 16 de outubro de 1864. [Côrte, 1864]. 12 pp. [*see* 251].

1817 d

Figueira de Mello, J. M. Chronica da rebellião praieira em 1848 e 1849. Rio de Janeiro, Typographia do Brasil de J. J. La Rocha, 1850. xv, 425 pp. *and* Documentos justificativos, 186 pp.

1818

Findlay, A. G. A sailing directory for the Ethiopic or southern Atlantic ocean; including the coasts of Brazil, the African coasts, etc.; fourth edition. London, Richard Holmes Laurie, 1855. xl, 578, 24 pp.

1819 a d

Fuentes, M. A. Lima: or, sketches of the capital of Peru . . . Paris, Firmin Didot, 1866. ix, 224 pp., illus., plates.

1820 a d

Gavet, D. and P. Boucher. Jakaré-Ouassou: ou les Tupinambas, chronique brésilienne. Paris, Timothée de Hay, 1830. 446 pp.

1821 a d h

Gerber, H. Noções geographicas e administrativas da provincia de Minas Geraes. Rio de Janeiro, Georges Leuzinger, 1863. [i], 86 pp., plate, map. Second copy bound in. *See also* translation by Burton (47).

1822 e

Great Britain. Parliament. Correspondence respecting hostilities in the River Plate in continuation of papers presented to parliament in 1867. London, Harrison, printer, 1868. [ii], 44 pp. At head of title: River Plate. No. 1 (1868). [*see* 250].

1823

— — Correspondence respecting hostilities in the River Plate (in continuation of papers presented to parliament, February 1868). London, Harrison, printer, 1868. iv, 40 pp., map. At head of title: River Plate. No. 2 (1868). [*see* 250].

1824 d l

Halfield, H. G. F. Relatorio concernente a exploracão do Rio de S. Francisco desde a Cachoeira da Pirapora Até o oceano atlantico durante os annos de 1852, 1853 e 1854. Rio de Janeiro, Typographia Moderna de Georges Bertrand, [1860], [iii], 216 pp.

1825

Hannah, J. Review of "Sheep farming in Buenos Ayres," etc. by Wilfrid Latham, Esq. Buenos Aires, Imprenta "Buenos Aires", 1867. 72 pp. [*see* p.].

1826

[—]. Sheep-husbandry in Buenos Ayres: a continuation of the discussion between Wilfrid Latham, Esq., and the author. Buenos Ayres, Imp. Buenos Aires, 1868. 65 pp. [*see* 223].

1827 d i j

Hartt, C. F. Notes on the Lingoa geral, or modern Tupi of the Amazonas. [Hartford, etc.], 1872. 20 pp. Reprinted from the *Transactions of the American Philological Association,* 1872. (Ch. Fred. Hartt, Ithaca, N.Y., 11 Aug. 1873). [*see* 255].

1828

— O mytho do Curupira. [Ithaca, N.Y.], 1873. 12 pp. Reprinted from *Aurora brasileira: Jornal dos Estudantes brasileiros no Universidade de Cornell,* Nos. 1–2. [*see* p.].

1829
Head, Sir F. B. Rough notes taken during some rapid journeys across the pampas and among the Andes . . . new edition. London, John Murray, 1861. ix, 166 pp.

1830 a d l
Herndon, W. L. Exploration of the valley of the Amazon, made under the direction of the Navy Department, by Wm. Lewis Herndon and Lardner Gibbon. Part 1, by Lieut. Herndon. Washington, Robert Armstrong, public printer, 1853. 414, iv pp., front., plates.

1831 d
Hinchliff, T. W. South American sketches: or a visit to Rio de Janeiro, the Organ mountains, La Plata, and the Paranà. London, Longman, Green, [etc.], 1863. xix, 414 pp., front., plates (3 col.), map.

1832
Homem de Mello, F. I. M., preface. Esboços biographicos. Rio de Janeiro, Typographia do Diario do Rio de Janeiro, 1862. 2 vols. (Bibliotheca brasileira, 2, 4). [*see* 252].

1833 a d
Humboldt, A. von. Personal narrative of travels to the equinoctial regions of America, during the years 1799–1804; translated and edited by Thomasina Ross. London, Henry G. Bohn, 1852–3. 3 vols. (Bohn's scientific library).

1834
Hunt, L. and others. Brazil: report . . . on the trade . . . for the year 1863. [Rio de Janeiro], 1863. 33–80 pp., tables. [*see* 255].

1835 d
Hutchinson, T. J. Buenos Ayres and Argentine gleanings: with extracts from a diary of Salado exploration in 1862 and 1863. London, Edward Stanford, 1865. xxiii, 321 pp., front. (port.), illus., plates, maps.

1836 d h j
— The Paraná, with incidents of the Paraguayan war, and South American recollections, from 1861 to 1868. London, Edward Stanford, 1868. xxvii, 424 pp., front. (port.), illus., plates, map. (John King, 22 Oct. 1871 to writer's mother).
k
Another copy.

1837 a d h
— Two years in Peru, with exploration of its antiquities. London, Sampson Low, Marston, [etc.], 1873. 2 vols., fronts., illus., plates, map.

1838
— Up the rivers and through some territories of the Rio de la Plata districts in South America. London, Bates Hendy, [1868]. 30 pp. [*see* 251].

1839 d
Imperial Collegio de Pedro II, Rio de Janeiro. Programma do ensino da instrucção secundario do Municipio da Corte, tanto no internato como no externato . . . Rio de Janeiro, Typographia Nacional, 1858. 37 pp. [*see* 253].

1840 a d h
Instituto Historia do Brazil. Historia geral do Brazil. Rio de Janeiro, E. e H. Laemmert, 1854–7. 2 vols., front., plates, maps.

1841
Lacerda e Almeido, F. J. de. Diario da viagem . . . pelas Capitanias do Parà, Rio Negro, Matto-Grosso, Cuyabà e S. Paulo, nos annos de 1780 a 1790. S. Paulo, Typ. de Costa Silveira, 1841. 90 pp. [*see* p.].

1842 a d
Las Casas, B. de. Relation des voyages et des de'couvertes que les Espagnols ont fait dans les Indes occidentales . . . Amsterdam, J. Louis de Lorme, Librairie sur le Rockin, 1698. [xii], 402 pp., front.

1843 e
— Oeuvres do Don Barthélemi de Las Casas . . . précédées de sa vie . . . par J.-A. Llorente. Paris, Alexis Eymery, 1822. 2 vols., front. (port.).

1844 d
Leitão da Cunha, P. Relatorio apresentado a'o exm. 1º Vice Presidente de provincia de Santa Catharina o commendador Francisco José d'Oliveira. Desterro, Typo Commercial, [1864]. 42 pp. [*see* 242].

1845 d
Lopez, V. F. Les races aryennes du Pérou: leur langue, leur religion, leur histoire. Paris, A. Franck; Montevideo, chez l'auteur, 1871. vi, 424 pp.

1846 a d

Luccock, J. Notes on Rio de Janeiro, and the southern parts of Brazil, taken during a residence of ten years in that country, from 1808 to 1818. London, Samuel Leigh, 1820. xvi, 640 pp., maps.

1847 a d

Macedo, J. M. de. Um passeio pela cicade do Rio de Janeiro. Rio de Janeiro, J. M. Nunes Garcia, Candido Augusto de Mello, 1862–3. 2 vols., plates.

1848

— Notions on the geography of Brazil; translated by H. Le Sage, Leipzig, F. A. Brockhaus, printer, 1873. vi, 576 pp., tables. [*see* 1901].

1849 a d h j

Machado d'Oliveira, J. J. Geographia da provincia de S. Paulo . . . [São Paulo, Typ. Imparcial de J. R. de A. Marques, 1862]. xiv, 122 pp. (A. T. Coimbra, Rio de Janeiro, n.d.).

1850 d i

— Quadro historico da provincia de São Paulo . . . São Paulo, Typographia Imparcial de J. R. A. Marques, 1864. 345 pp. (author to Senr. Garraux).

1851 a d

Magalhâes, C. de. Região e raças, selvagens do Brasil . . . Rio de Janeiro, Typ. de Pinheiro, 1874. 160 pp. At head of title: Ensaio de anthropologia. [*see* 255].

1852 a d

Martin de Moussy, V. Description géographique et statistique de la Confédération Argentine. Paris, Firmin Didot, 1860–4. 3 vols.

1853 a d

Mawe, J. Voyages dans l'intérieur du Brésil, particulièrement dans les districts de l'or et du diamant; traduits de l'anglais par J.-B.-B. Eyriès. Paris, Gide Fils, 1816. 2 vols., fronts., plates.

1854

Maynard, F. Voyages et aventures au Chili. Paris, Librairie Nouvelle, Jaccottet, Bourdilliat, 1858. [iii], 308 pp.

1855 a

Meiggs, E. Reseña histórica del ferrocarril entre Santiago i Valparaiso. Santiago, Imprenta del Ferrocarril, 1863. [iv], 173 pp., front. (port.), plates, map. [*see* 248].

1856

Mello Moraes, A. J. de. Brasil historico escripto . . . 2a serie – 1866-1867. Rio de Janeiro, Pinheiro; Fauchon & Dupont, 1866–7. 2 vols., illus. [*see* q.].

1857

Molina, J. I. Compendio de la historia geografica, natural y civil del reyno de Chile, escrito en Italiano, traducida en Español por Don Domingo Joseph de Arquellada Mendoza. Madrid, Antonio de Sancha, 1788–95. 2 vols., front., maps, tables.

1858 l

Moreira de Barros, A. Relatorio apresentado á assembléa legislativa provincial das Alagôas. Maceió, Typographia do Jornal-O Progressista, 1867. [ii], 19 pp. [*see* 250].

1859 a d

Moure, A. Les Indiens de la province de Mato-Grosso (Brésil): observations. Paris, 1862. 56 pp. Reprinted from *Nouvelles annales des voyages*, April, June, July 1862. [*see* 254].

1860 d

Mulhall, M. G. The cotton fields of Paraguay and Corrientes . . . Buenos Ayres, M. G. and E. T. Mulhall, 1864. 120 pp. [*see* 251].

1861

— and E. T. Mulhall. Handbook of the river Plate; comprising Buenos Ayres, the upper provinces, Banda Oriental, and Paraguay. Vol. 1. Buenos Ayres, Standard Printing Office, 1869. xvi, 192, 160, 200 pp.

1862 a

[Nunez, I.]. An account, historical, political, and statistical, of the United provinces of Rio de la Plata; with an appendix, concerning the usurpation of Monte Video by the Portuguese and Brazilian governments; translated from the Spanish. London, R. Ackermann, 1825. x, 345 pp., front. (map), map.

1863 d

Oran, pseud. Panama. New York, 1859 [20 pp.]. From *Harper's new monthly magazine*, Sept. 1859, Vol. 19, No. 112, pp. [433]–454. [*see* 246].

1864 h

Ottoni, C. B. Esboço historico das estradas de ferro do Brazil. Rio de Janeiro,

J. Villeneuve, 1866. 30 pp. [*see* 250].
Another copy. [*see* 250].

1865 a d
[Palivisino, V.]. Memoria sobre la Araucania por un misionero del Colejio de Chillan. Santiago de Chile, Imprenta de la Opinion, 1860. iv, 166 pp. [*see* 507].

1866
[Paraguay. On Paraguay. Besanzon, Impr. de J. Jacquin, after 1857]. 129–364 pp. Contents: Chapters 3–7, Geografia, hidrografia, poblacion, tribus, Indias, lengua guarani. wanting title-page etc. [*see* 254].

1867 d
Paris International Exhibition. Catalogue of the articles sent to the Universal Exhibition at Paris in 1867. [Rio de Janeiro, E. & H. Laemmert, 1867]. [v], 197 pp. [*see* 255].

1868 a d
— The empire of Brazil at the . . . exhibition of 1867. Rio de Janeiro, E. & H. Laemmert, 1867. 139 pp. [*see* 255].

1869
Pascual, A. D. de. As quatro derradeiras noites dos inconfidentes de Minas Geraes (1792). Rio de Janeiro, Typographia do Imperial Instituto artistico, 1868. [6], x, 138 pp. At head of title: Um epysodio de historia patria. [*see* 254].

1870 i
— Ensaio critico sobre a viagem ao Brasil em 1852 de Carlos B. Mansfield. Rio de Janeiro, Typographia Universal de Laemmert, 1861–2. 2 vols. (in 1).

1871 a d
Pereira de Vasconcellos, J. M. Ensaio sobre a historia e estatistica da provincia do Espirito Santo. Victoria, P. A. d'Azeredo, 1858. 254 pp., front. [*see* 242].

1872 a d
Perez-Rosales, V. Essai sur le Chili. Hambourg, Imprimé chez F. H. Nestler & Melle, 1857. xxiii, 455 pp., plate, map, tables. [*see* 507].

1873 d
Perkins, G. Relacion de la espedicion á el rey en el Chaco. Rosario, El Ferro-Carril, 1867. 78, x pp., map. [*see* 249].

1874 a
Pim, B. The Negro and Jamaica; read before the Anthropological Society of London, February 1, 1866. London, Trübner, 1866. vii, 72 pp. Upper wrapper at head: Special number of 'The popular magazine of anthropology'. [*see* p.].

1875
Pinto Junior, J. A. O charlatão Carlos Expilly e A verdade sobre o conflicto entre o Brasil, Buenos-Ayres, Montevidéo e o Paraguay. S. Paulo, Typ. Allemã de H. Schroeder, 1866. 22 pp. [*see* 252].

1876 i
— Uma excursão a'comarca de Iguape. S. Paulo, Typ. americana, 1866. 77 pp. [*see* 252].

1877 a d
Rath, C. Fragmentos geologicos e geographicos etc. . . . de S. Paulo e Paraná . . . começados no anno de 1845. S. Paulo, Typographia Imparcial, 1856. [iii], 78 pp. [*see* 242].

1878 d i
Rickard, F. I. The mineral and other resources of the Argentine Republic (La Plata) in 1869. London, Longmans, Green, 1870. 323 pp. (March 1871).

1879 a d
The River Plate hand-book, guide, directory and almanac for 1863 . . . First year. Buenos Ayres, Editors of the Standard, 1863. iii, 298 pp.

1880 a d
Rivero, M. E. and J. J. von Tschudi. Peruvian antiquities . . . translated into English from the original Spanish, by Francis L. Hawks. New York, George P. Putnam, 1853. xxii, 306 pp., front., illus.

1881
Robertson, J. P. and W. P. Robertson. Letters on South America; comprising travels on the banks of the Paraná and Rio de La Plata. London, John Murray, 1843. 3 vols.

1882 d
Rodriques, J. A. Apontamentos da populacaõ, topographia e noticias chronologicas do municipio da cidade de S. João del-Rei, provincia de Minas-Geraes. S. João d'El-Rei, Typ. de J. A. Rodriques, 1858. 27, ix pp. [*see* 251].

1883 i

Sada, L. Instituto y hacienda normal para la enseñanza de la agricultura de la República del Perù en Lima. Lima, Imprenta del Estado, 1870. 171 pp., front., plans.

1884 a d

Saint-Hilaire, A. de. Voyage dans les provinces de Rio de Janeiro et de Minas Geraes. Paris, Grimbert et Dorez, 1830. 2 vols., front. Half-title: Voyages dans l'intérieur du Brésil, Première partie.

1885

— Voyage dans le district des diamans et sur le littoral du Brésil. Paris, Gide, 1833. 2 vols. Half-title: Voyages dans l'intérieur du Brésil, Seconde partie.

1886 d

— Voyage aux sources du Rio de S. Francisco et dans la province de Goyaz. Paris, Arthus Bertrand, 1847–8. 2 vols. Half-title: Voyages dans l'intérieur du Brésil, Troisième partie.

1887 a d

— Voyages dans les provinces de Saint-Paul et de Sainte-Catherine. Paris, Arthus Bertrand, 1851. 2 vols. Half-title: Voyages dans l'intérieur du Brésil, Quatrième partie.

1888 i

Santa-Anna Nery, F. J. de. Le Brésil en 1889 . . . publíe par les soins du syndicat du Comité franco-brésilien pour l'Exposition universelle de Paris . . . Paris, Charles Delagrave, 1889. xix, 699 pp., map, charts. (Eduardo Prado, Paris, 26 Sept. 1889).

1889 d h

São Paulo. Provincia. Ensaio d'um quadro estatistico da provincia de S. Paulo. S. Paulo, Typographia de Costa Silveira, 1838. [8], iv, 86 pp., tables.

1890 a

— — Assembléa Legislativa. Diario da viagem do Dr. Francisco José de Lacerda e Almeida pelas do Pará, Rio Negro, Matto-Grosso, Cuyabá, e S. Paulo, nos annos de 1780 a 1790. S. Paulo, Typ. de Conta Silveira, 1841. 90 pp. [*see* 242].

1891 d i

Scully, W. Brazil: its provinces and chief cities: the manners & customs of the people, agricultural, commercial, and other statistics. London, Murray, 1866. xv, 398 pp., front. (map). (1 May 1866).

1892

Silva Mendes Leala Junior, J. da. Calabar: historia brasileira do seculo XVII. Rio de Janeiro, Typ. do Correio Mercantil, 1863. 4 vols. (in 1).

1893 a d

Southey, R. History of Brazil. London, Longman, Hurst, Rees, [etc.], 1810–19. 3 vols. Vol. 1, second edition, dated 1822.

1894 a d

Souza Brasil, T. P. de. Compendio elementar de geographia geral e especial do Brasil . . . quarto edição. Rio de Janeiro, Eduardo & Henrique Laemmert, 1864. viii, 556 pp.

1895 a h

Tavares Bastos, A. C. O valle do Amazonas: estudo sobre a livre navegação do Amazonas. Rio de Janeiro, B. L. Garnier, 1866. xxiii, 371 pp.

1896 l

Temple, E. Travels in various parts of Peru, including a year's residence in Potosi. London, Henry Colburn and Richard Bentley, 1830. 2 vols., fronts., plates, map.

1897 a d

Thompson, G. The war in Paraguay, with a historical sketch of the country and its people, and notes on the military engineering of the war. London, Longmans, Green, 1869. xi, 347 pp., front. (port.), maps.

1898

Uricoechea, E. Mapoteca colombiana: coleccion de los títulos de todos los mapas, planos, vistas etc. relativos á la América española, Brasil e islas adyacentes. Londres, Trübner, 1860. xvi, 215 pp.

1899 d i

Vasconcellos Menezes de Drummond, A. Memoria historica academica apresentada á Congregação dos Lentes da Facultade de Direito do Recife na sessão de 15 de março de 1864. Pernambuco, Typographia de Figueiroa de Faria & Filho, 1864. 104, vi pp., tables. (Recife, 19 July 1865). [*see* 249].

1900 d

Vieira Couto, J. Memoria sobre as minas da capitania de Minas Geraes. Rio de Janeiro, Eduardo e Henrique Laemmert, 1842. [viii], 160 pp. [*see* 242].

1901
Vienna Universal Exhibition. The empire of Brazil at the Vienna Universal Exhibition of 1873. Rio de Janeiro, E. & H. Laemmert, printer, 1873. 388 pp., maps, table.

1902 a d
Walsh, R. Notices of Brazil in 1828 and 1829. London, Frederick Westley and A. H. Davis, 1830. 2 vols., fronts. (maps), illus., plates, scores.

1903 d h
Watson, R. G. Spanish and Portuguese South America during the colonial period. London, Trübner, 1884. 2 vols., plate, map (in pocket).

1904 i
Wucherer, Ò. [On the ophidians of Bahia. London], 1861–3. [11] pp. From the *Proceedings of the Zoological Society of London,* 12 Nov. 1861 and 27 Jan. 1863. [*see* 225].

1905 a d
Yves, d'Évreux, Capuchin [Simon Michellet]. Voyage dans le nord du Brésil fait durant les années 1613 et 1614 par Père Yves d'Évreux . . . introduction et des notes par M. Ferdinand Denis. Leipzig & Paris, A. Franck, 1864. xlvi, 456, x pp.

1906-2098 Asia

1906 a d
Anonymous. Mahableshwur guide; compiled and arranged by an old frequenter of the hills. Bombay, Education Society's Press, Byculla, 1875. [ii], 35 pp., map. [*see* p.].

1907
— Turkish proverbs; translated into English. Venice, Armenian Monastery of S. Lazarus, printer, 1860. 35 pp. [*see* 260].

1908 a d h
'Abd al-Rahmān ibn Abi Bakr (Jalāl al-Dīn) al-Suyūti. History of the caliphs, by Jalálu'ddín A's Suyúti; translated from the original Arabic, by H. S. Jarrett. Calcutta, Asiatic Society of Bengal, 1881. xxiii, 563 pp. (Bibliotheca indica, Vol. 87).

1909 a d h
Aberigh-Mackay, G. R., compiler. Handbook of Hindustan. Bombay, The Times of India Office, 1875. x, 225, xviii pp., map. At head of title: The Prince's guide book.

1910 d i j
Abreu, G. de V. Buddhist legends: from "Fragmentos d'uma tentavia de estudo scoliastico do epopeia portugueza"; translated . . . by Donald Ferguson, Ceylon. [Bombay, 1884]. 56 pp. Reprinted from the *Indian antiquary,* Vol. 13. (Donald Ferguson, Colombo, 10 Oct. 1884). [*see* p.].

1911 a d
Abreu, M. V. d'. O governo do Vice-Rei Conde do Rio Pardo no estado da India portugueza, desde 1816 ate' 1821: memoria historica. Nova-Goa, Imprensa-Nacional, 1869. [6], iv, 253, 3 pp.

1912 d
Ainsworth, W. F. The Euphrates valley railway. London, Adams & Francis, 1872. 70 pp., front. (map). [*see* 258].

1913 d
— Travels and researches in Asia Minor, Mesopotamia, Chaldea, and Armenia. London, John W. Parker, 1842. 2 vols., fronts., illus., maps.

1914
Aissé Chanum Effendi. Les malheurs et la fuite de Aissée Chanum Effendi, fille de son Altesse Kibrizli-Mehemet. Paris, Georges Kugelmann, 1870. 29 pp. [*see* 257].

1915 a d h i
Ali ibn Husain, al-Mas'ūdī. Les prairies d'or; texte et traduction par C. Barbier de Meynard et Pavet de Courteille. Paris, Imprimerie Impériale, 1861–77. 9 vols. (W. G. Palgrave).

1916
Andrew, Sir W. P. The advance of Russia: being a letter published in "The Times" in August last; with an appendix by A. Sprenger. London, W. H. Allen, 1886. vii, 18 pp. [*see* p.].

1917 a e
— The Euphrates route to India: letters addressed to the British and Turkish governments, etc. London, Wm. H. Allen, 1871. 38 pp. [*see* 258].

1918
— The Euphrates valley railway: letters addressed to Her Majesty's Secretaries of State for Foreign Affairs, and for India. London, Wm. H. Allen, 1870. 53 pp., front. (map). [*see* 258].

1919 a d h i
— Memoir on the Euphrates valley route to India . . . correspondence and maps.

London, Wm. H. Allen, 1857. xvi, 267 pp., front. (map), map. (19 July 1861).
1920 e
— On the completion of the railway system of the valley of the Indus: a letter to His Grace the Duke of Argyll, K.T. (Secretary of State for India in Council). London, Wm. H. Allen, 1869. [iii], 124 pp., maps. [*see* 256].
Another copy. [*see* 258].
1921 d
Albert de Luynes, H. d'. Mémoire sur le sarcophage et l'inscription funéraire d'Esmunazar, Roi de Sidon. Paris, Henri Plon, 1856. vi, 83 pp., plates. [*see* q.].
1922 a d
Barros, J. de. Da Asia . . . dos feitas, que os Portuguezes fizeram no descubrimento, e conquista dos mares, e terras do Oriente [and] Vida de João de Barros por Manoel Severim de Faria e Indice geral das quatro decadas da sua Asia. Lisboa, Regia Officina Typografica, 1777–8. 4 vols. (in 8) *and* Vida. wanting Vol. 1, no. 1, Vol. 3, no. 1. *See also* 1946.
1923 d h
Bâsim, the Smith. Bâsim le forgeron et Hârûn er-Rachîd; texte arabe en dialecte d'Egypte et de Syrie; publié . . . et accompagné d'une traduction et d'un glossaire, par le Comte Carlo de Landberg. 1, Texte, traduction et proverbes. Leyde, E. J. Brill, 1888. xvii, 87, [116] pp.
1924 e
[Behm, E.]. Review of The land of Midian revisited. Gotha, 1879]. 156–7 pp. From *Petermann's geographische Mittheilungen,* Vol. 25, No. 4. [*see* p.].
1925
Beke, C. T. Discoveries of Sinai in Arabia and of Midian; edited by his widow. London, Trübner, 1878. xix, 606 pp., front. (port.), plates, map.
1926 i l
— A few words with Bishop Colenso on the subject of the exodus of the Israelites and the position of Mount Sinai; third edition. London, William & Norgate, 1863. 14 pp. [*see* 241].
1927
— The idol in Horeb: evidence that the golden image at Mount Sinai was a cone, and not a calf. London, Tinsley, 1871. vii, 155 pp.
1928 i
— Mount Sinai, a volcano. London, Tinsley, 1873. 48 pp.
1929 a d
Beke, Mrs E. Jacob's flight: or a pilgrimage to Harran. London, Longman, Green, [etc.], 1865. xii, 360 pp., front., plates, map.
1930 a i
Bell, E. The Bengal reversion, another "exceptional case.". London, Trübner, 1872. xxxix, 83 pp.
1931 d h i
Bellasis, A. F. An account of the hill station of Matharan, near Bombay. Bombay, Education Society's Press, Byculla, 1869. 28 pp. (Dr Niven, Supt of Matharan, Feb. 1876).
1932 d i l
Besant, W. and E. H. Palmer. Jerusalem: the city of Herod and Saladin. London, Richard Bentley, 1871. viii, 492 pp., front.
1933 d
British Syrian Schools. Ladies' Association for the Social and Religious Improvement of the Syrian Females; eighth annual report, 1868. London, Seeley, Jackson & Halliday, 1869. 134 pp., front., map.
1934 d
Brown, J. P. The dervishes: or oriental spiritualism. London, Trübner, 1868. viii, 415 pp.
1935 a d i
Burnell, A. C. A tentative list of books and some MSS. relating to the history of the Portuguese in India proper. Mangalore, Basel Mission Press, printer, 1880. vi, 133 pp.
1936 d
Cameron, V. L. Our future highway to India. London, Macmillan, 1880. 2 vols., front., plates, map.
1937 d h i
Casartelli, L. C. La philosophie religieuse du Mazdéisme sous les Sassanides. Paris, Maisonneuve Frères et Ch. Leclerc, [etc.], 1884. viii, 192 pp. (3 Feb. 1887).

1938 a d

Chesney, F. R. Narrative of the Euphrates expedition . . . during the years 1835, 1836 and 1837. London, Longmans, Green, 1868. xviii, 564 pp.

1939

— Report of the Euphrates valley railway. London, Smith, Elder, 1857. [ii], 20 pp. [*see* 256].

1940 d h l

Churchill, C. H. Mount Lebanon: a ten years' residence from 1842 to 1852, describing the manners, customs, and religion of its inhabitants. London, Saunders & Otley, 1853. 3 vols., fronts., plates, map.

1941 d

Clarke, H. On public instruction in Turkey. [London], 1867. 502–34 pp., tables. Reprinted from the *Journal of the Statistical Society,* Dec. 1867. [*see* 256].

1942 d

— On the supposed extinction of the Turks and increase of the Christians in Turkey . . . Constantinople, Koehler, 1865. [viii], 261–93 pp., tables. Reprinted from the *Journal of the Statistical Society,* June 1865. [*see* 256].

1943 a d

Clermont-Ganneau, C. Trois inscriptions de la Xe Légion fretensis trouvées à Jérusalem. Paris, Imprimerie nationale, 1872. 16 pp., illus. [*see* 259].

1944 a e

Cooke, G. W. China: being "The Times" special correspondence from China in the years 1857–58 . . . with corrections and additions by the author. London, G. Routledge, 1858. xxxii, 457 pp., front. (port.), plate, maps (1 in pocket).

1945 d h i

Cottineau de Kloguen, D. L. An historical sketch of Goa, the metropolis of the Portuguese settlements in India. Madras, William Twigg, printer, 1831. [vi], 188 pp., front. (map). (Owen Collington, Feb. 1876).

1946 a d l

Couto, D. de. Da Asia . . . dos feitas, que os Portuguezes fizeram na conquista, e descubrimento das terras e mares do Oriente. Lisboa, Regia Officina Typographica, 1778–88. 8 vols. (in 14), *and* Indice geral das decadas de Couto. Half-title: Continuacão Da Aśia de João de Barros. *See also* 1922.

1947 a d

Crichton, A. History of Arabia, ancient and modern; second edition. Edinburgh, Oliver & Boyd, 1834. 2 vols., plates.

1948 a d

Dabistan. The Dabistan: or school of manners; translated from the original Persian, with notes and illustrations, by David Shea . . . and Anthony Troyer; edited with a preliminary discourse by the latter. Paris, printed for the Oriental Translation Fund of Great Britain & Ireland, 1843. 3 vols.

1949 a d h

Davis, J. F. The Chinese: a general description of the Empire of China and its inhabitants. London, Charles Knight, 1836. 2 vols., illus. wanting Vol. 1.

1950 a d l

Dellon, C. Narraçao da inquisicão de Goa; escripta em Francez; vertida em Portuguez . . . por Miguel Vicente d'Abreu. Nova-Goa, Imprensa Nacional, 1866. x, 310 pp.

1951

Dethier, P. A. Le Bosphore et Constantinople: description topographique et historique. Vienne, Alfrede Hölder, 1873. 79 pp., front. (map), illus., plate. [*see* 230].

1952 a

Didier, C. Séjour chez le Grand-Chérif de la Mekke. Paris, L. Hachette, 1857. vii, 311 pp.

1953 k l

Dixon, W. H. The Holy Land; copyright edition. Leipzig, Bernhard Tauchnitz, 1865. 2 vols. (in 1). Lettered on upper cover: Holy Land / Isabel Burton.

1954 a d h i j

Doughty, C. M. Travels in Arabia deserta. Cambridge, University Press, 1888. 2 vols., plates, maps (in pocket). (Ch. M. Doughty, Mattock, 27 Jan. 1885; and 3 cards, 21 Oct. and 13 Nov. [1884], 10 Jan. [1888]).

1955 a d h

Dozy, R. De Israëlieten te Mekka van Davids tijd tot in de vijfde eeuw onzer tijdrekening. Haarlem, A. C. Kruseman, 1864. vi, 214 pp., plate.

1956 a d h

Drake, C. F. T. The literary remains of the late Charles F. Tyrwhitt Drake; edited with a memoir by Walter Besant. London, Richard Bentley, 1877. [vii], 320 pp., front. (port.).

1957

Du Couret, L. Les mystères du désert: souvenirs de voyages en Asie et en Afrique, par Hadji-Abd' el-Hamid-Bey (Cel. L. du Couret). Paris, E. Dentu, 1859. 2 vols. (in 1), maps.

1958 d j

Duff, J. C. D. G. History of the Marat,has; translated from the English original of Captain Grant Duff, by Captain David Capon; [Marathi text]. Bombay, 1830. 6, 29, 416, [4] pp. (Bapoojee Hurree, Booldanah, 9, 15, 29 Feb., March and 14 March 1876).

1959 a

Eastern Question Association. Report of the proceedings of the National Conference at St. James's Hall, London, December 8th, 1876. London, James Clarke, [1876]. xv, 136 pp. [*see* p.].

1960 d

Eastwick, E. B. Gold in India: whence the gold of India and is it exhausted? London, Chapman & Hall, 1880. 35 pp. Reprinted from the *Gentleman's Magazine,* Vol. 246. [*see* p.].

1961 d

Elliot, G. Indian remounts: being letters to the late Earl of Mayo, and the Right Honour[ab]le Lord Northbrook. Bombay, "Bombay Gazette" Steam Press, 1874. 19 pp. [*see* p.].

1962 d h

Elphinstone, M. The history of India. London, John Murray, 1841. 2 vols., map.

1963 d

Fergusson, J. The Holy sepulchre and the temple at Jerusalem. London, John Murray, 1865. xvi, 151 pp., front.

1964

Fisk, G. A pastor's memorial of Egypt, the Red sea, the wildernesses of Sin and Paran, Mount Sinai, Jerusalem, and other principal localities of the Holy Land; fifth edition. London, Hamilton, Adams, 1853. viii, 469 pp., front., plates.

1965 d

Freshfield, D. W. Travels in the central Caucasus and Bashan, including visits to Ararat and Tabreez and ascents of Kazbek and Elbruz. London, Longmans, Green, 1869. xv, 509 pp., front., illus., plates, maps.

1966 k l

Fullerton, A. F. A lady's ride through Palestine & Syria, with notices of Egypt and the canal of Suez. London, S. W. Partridge, 1872. viii, 349 pp., front., plates.

1967

Gavazzi, M. Alcune notizie raccolte in un viaggio a Bucara. Milano, La Perseveranza, 1865. 172 pp., map. [*see* 258].

1968 k

Geary, G. Through Asiatic Turkey: narrative of a journey from Bombay to the Bosphorus. London, Sampson Low, Marston, [etc.], 1878. 2 vols., fronts., illus., plates, maps.

1969 d

Gerson da Cunha, J. Contributions to the study of Indo-Portuguese numismatics. Bombay, Education Society's Press, 1880. 19–34 pp., plate. Fasciculus II. [*see* p.].

1970 a d j

— The English and their monuments at Goa. Bombay, Education Society's Press, Byculla, 1877. 28, iv pp. (J. Gerson da Cunha, Bombay, 30 May 1877). [*see* p.].

1971 a d i

— An historical and archaeological sketch of the island of Angediva. [Bombay], 1875. 24 pp. Reprinted from the *Journal of the Bombay Branch of the Royal Asiatic Society,* Vol. 11. [*see* p.].

1972 d i

— Memoir on the history of the tooth-relic of Ceylon; with a preliminary essay on the life and system of Gautama Buddha. London, W. Thacker, [etc.], 1875. xiii, 71 pp., front., plates.

1973 a d

— Notes on the history and antiquities of Chaul and Bassein. Bombay, Thacker,

Vining, 1876. xvi, 263 pp., front., plates, map.

1974 a d

Ghazâlî, Aboû Hâmid Mohammed ibn Mohammed al-. Ad-dourra al-Fâkhira: la perle précieuse de Ghazâlî, traité d'eschatologie musulmane . . . avec une traduction française, par Lucien Gautier. Genève-Bäle, Lyon, H. Georg, [etc.], 1878. xvi, 90, [114] pp.

1975 d

[Gillmore, P.]. The black pamphlet: the famine of 1874, by Ubique. Calcutta, W. Newman, 1875. 60 pp. [*see* p.].

1976

Goa. Almanach de Goa para o anno bissexto de 1876, por Mariano José Corrêa da Silva. 1 sheet. [*see* p.].

1977 d

Gomes, F. L. A liberdade da terra e a economia rural da India portugueza. Lisboa, Typographia Universal, 1862. 102 pp. [*see* 249].

1978

Gordon, C. G. General Gordon's private diary of his exploits in China; amplified by Samuel Mossman. London, Sampson Low, Martson, [etc.], 1885. xv, 302 pp., front. (port.), plate, map.

1979 l

Great Britain. Rules of Her Britannic Majesty's supreme consular court and other consular courts in the dominions of the Sublime Ottoman Porte. London, Her Majesty's Stationery Office, 1863. 118 pp. [*see* 257].

1980

— Parliament. Convention of commerce and navigation between Her Majesty and the Sultan of the Ottoman Empire . . . signed at Balta-Liman, near Constaninople, August 16, 1838. [n.p.], 1838. 8, 24 pp. [*see* 222].

1981 d

— — East India (Mirza Ali Akbar) . . . copy of further correspondence respecting Mirza Ali Akbar . . . ordered by the House of Commons to be printed. [London], 1860–1. 3 pts. [*see* p.].

1982 a d

Guarmani, C. C. El Kamsa: il cavallo arabo puro sangue: studio di venti anni in Siria Palestina egitto nei deserti dell'Arabia e nel Neged; seconda edizione. Gerusalemme, Tipografia dei PP. Francescani, 1866. 215 pp.

1983

— Il Neged settentrionale . . . itinerario da gerusa lemme a Aneizeh nel Cassim. Gerusalemme, Tipografia dei PP. Francescani, 1866. 210, xxxvii pp., map.

1984

Guys, H. La nation druse: son histoire, sa religion, ses moeurs et son état politique. Paris, chez France, 1863. 248 pp. [*see* 240].

1985

Haines, C. R. A vindication of England's policy with regard to the opium trade. London, W. H. Allen, 1884. [viii], 140 pp.

1986

Hamilton, C. Oriental zigzag: or wanderings in Syria, Moab, Abyssinia, and Egypt. London, Chapman & Hall, 1875. viii, 304 pp., front. (port.), plates.

1987 a d

Hamilton, W. J. Researches in Asia Minor, Pontus, and Armenia; with some account of their antiquities and geology. London, John Murray, 1842. 2 vols., front., plates, map.

1988

Harcourt, A. F. P. Our northern frontier: being observations on the recent advances of Russia towards Hindoostan. London, James Madden, 1869. 27 pp. [*see* 222].

1989 a

Herbelot, B. de. Bibliothèque orientale: ou dictionnaire universel. Maestricht, J. E. Dufour & Ph. Roux, 1776. 954 pp. [*see* q.].

1990 a d

Huggins, W. Sketches in India, treating on subjects connected with the government. London, John Letts, Jun., 1824. vi, 237 pp.

1991 a d

Hughes, A. W., compiler. A gazetteer of the Province of Sindh. London, George Bell, 1874. viii, 898 pp., plate, maps (1 in pocket).

1992 a d h

Hunter, F. M., compiler. The Aden handbook: a summary of useful information regarding the settlement. London, Harrison, 1873. iv, 72 pp. [*see* p.].

1993 a d

Ibn Batuta. The travels of Ibn Batuta; translated from the abridged Arabic manuscript copies . . . in the Public Library of Cambridge; with notes . . . by Samuel Lee. London, printed for the Oriental Translation Committee, 1829. xix, 243 pp.

1994

Imray, J. F., compiler. Sailing directions for the Levant, or eastern portion of the Mediterranean sea; compiled from the latest surveys. London, James Imray, 1866. [iii], 40 pp. [*see* p.].

1995

[India Office. Papers on the Euphrates railway. London, 1867]. [24] pp. [*see* 256].

1996 d

Irving, W. The life of Mahomet, [and] Lives of the successors of Mahomet. London, Henry G. Bohn, 1850. 2 vols. (in 1), front. (Bohn's standard library).

1997

— — London, George Bell, 1876–7. 2 vols. (in 1), front. (Bohn's standard library).

1998 a d

Ja'far Sharif. Qanoon-e-Islam: or the customs of the Mussulmans of India, comprising a full and exact account of their various rites and ceremonies, from the moment of birth till the hour of death, by Jaffur Shurreef; composed under the direction of, and translated by G. A. Herklots . . . second edition, carefully revised. Madras, J. Higginbotham, 1863. xxiii, 296, cxliii pp., 2 fronts., plates.

1999

Kautzsch, E. Die Siloahinschrift. [Leipzig], 1881. 260–72 pp., plate. Reprinted from *Zeitschrift des deutschen Palästina-Vereins,* Vol. 4. [*see* p.].

2000 a d

— and A. Socin. Die Aechtheit der moabitischen Alterthümer. Strassburg, Karl J. Trübner; London, Trübner, 1876. viii, 192 pp., plates.

2001 d i j

Keane, J. F. My journey to Medinah. London, Tinsley Brothers, 1881. viii, 212 pp. ([London], 3 Oct. 1881).

2002 a d h j

— Six months in Meccah. London, Tinsley Brothers, 1881. [x], 212 pp. ([London], 2 and 10 Aug. 1881).

2003 d l

Kent, S. H. Gath to the Cedars: experiences of travel in the Holy Land and Palmyra, during 1872. London, Frederick Warne, [etc., 1874]. xv, 373 pp., front., plates.

2004 d l

[Kinglake, A. W.]. Eothen; fourth edition. London, John Ollivier, 1845. xi, 423 pp., front., plate.

2005

Kinloch, A. The Andaman islands, their colonization, &c.: a correspondence addressed to the India Office. London, Robert John Bush, 1870. 16 pp. [*see* p.].

2006 e

Kremer, A. von. Ueber dem shî'itischen Dichter Abu-lkâsim Mohammed ibn Hâni. [Leipzig, 1870]. 481–94 pp. Reprinted from *Zeitschrift der Deutschen morgenländischen Gesellschaft,* Vol. 24. [*see* 258].

2007 l

Kurrachee. Municipal Library and Museum. Catalogue of the Kurrachee Municipal Museum, established 1851. Kurrachee, printed at the "Commerical Press", 1874. [i], 80 pp. [*see* p.].

2008 a d

— — The twenty-second annual report of the Managing Committee of the Kurrachee Municipal Library and Museum, for the year ending 30th September 1874. Kurrachee, printed at the "Commerical Press", 1874. [i], 23 pp., plates. [*see* p.].

2009 a d

Landberg, C. de, ed. Critica arabica. Leiden, E. J. Brill, 1886–8. 2 pts. Part 1, wrapper dated 1887. [*see* p.].

2010 a d g h j

— Proverbes et dictions de la Province de Syrie, section de Saydâ. Leide, E. J. Brill; Paris, Maisonneuve, 1883. lii, 458, [6] pp. (Chas. Clarke, Gajazig, 24 Sept. 1886).

2011 d

Lenormant, F. Chaldean magic: its origin and development; translated from the French. London, Samuel Bagster, [1877]. xiii, 414, 18 pp.

2012 d
Liéven de Hamme, Franciscan. Guide indicateur des sanctuaires et lieux historiques de la Terre-Sainte. Jérusalem, Imprimerie des PP. Franciscains, 1869. xv, 700 pp.
d k l
Another copy.
2013 i
Loth, O. A catalogue of the Arabic manuscripts in the library of the India Office. London, printed by order of the Secretary of State for India in Council, 1877. viii, 324 pp.
2014 a d h j
Low, C. R. History of the Indian navy (1613–1863). London, Richard Bentley, 1877. 2 vols. (C. R. Low, [London], 25 June 1878).
2015
Mackintosh, Mrs. Damascus and its people: sketches of modern life in Syria. London, Seeley, Jackson & Halliday, 1883. xii, 296 pp., front., illus.
2016 d i l
Maclean, J. M., ed. Maclean's guide to Bombay. Bombay, "Bombay Gazette" Steam Press, 1875. v, 260, [ii], 32, xi, 50, ix pp., front. (maps).
2017 d l
Manu. Manava-dherma-sástra: or the institutes of Menu; edited by Graves Chamney Haughton. London, Cox & Baylis, printer; Rivingtons & Cochran, 1825. 2 vols. Contents: 1, Sanscrit text – 2, translated, with a preface, by Sir William Jones.
2018 d h i
Manuel, King of Portugal. The Italian version of a letter from the King of Portugal (Dom Manuel) to the King of Castille (Ferdinand), written in 1505, giving an account of the voyages to and conquests in the East Indies from 1500 to 1505 A.D.; reprinted from the copy . . . in the Marciana library at Venice . . . by A. C. Burnell. London, Wyman, printer, 1881. viii, 24 pp.
2019 d l
Martin, R. M. China: political, commercial, and social. London, James Madden, 1847. 2 vols., front. (map), illus., maps.
2020
Marvin, C. The railway race to Herat: an account of the Russian railway to Herat and India. London, W. H. Allen, 1885. 32 pp., illus., map. At head of title: Distribution edition. [*see* p.].
2021 d l
[Maurice, T.]. Brahmanical fraud detected: or the attempts of the sacerdotal tribe of India to invest their fabulous deities and heroes with the honours and attributes of the Christian Messiah, examined, by the author of Indian antiquities. London, printed for the author by W. Bulmer, 1812. viii, 144 pp. [*see* 1079].
2022 d i j
Miers, E. J. On a small collection of crustacea made by Major Burton in the gulf of Akaba. [London], 1878. 406–11 pp. Reprinted from *Annals and magazine of natural history*, Nov. 1878. (Edward J. Miers, [London], 16 Nov. 1878). [*see* p.].
2023 e
Millingen, F. Wild life among the Koords. London, Hurst & Blackett, 1870. xiii, 380 pp., front., map.
2024 i
Milne, J. Geological notes on the Sinaitic peninsula and north-western Arabia. [London], 1875. 28 pp., table. Reprinted from *Quarterly journal of the Geological Society of London*, Vol. 31. [*see* p.].
2025 a d i
Montet, E. Essai sur les origines des partis Saducéen et Pharisien et leur histoire jusqu'à la naissance de Jésus-Christ. Paris, Fischbacher, 1883. xvi, 334 pp.
2026
— La religion et le théatre en Perse. Paris, Ernest Leroux, 1887. 16 pp.
2027 d
Much, M. Ueber die Priorität des Eisens oder der Bronze in Ostasien. [Vienna, 1880]. 5 pp. Reprinted from *Mittheilungen der anthropologischen Gesellschaft in Wien*, Vol. 9. [*see* p.].
2028 d i
Muhammad ibn Ahmad, al-Mukaddasi. Description of Syria, including Palestine, by Mukaddasi; translated from the Arabic and annotated by Guy Le Strange. London, Palestine Pilgrims' Text Society, 1886. xvi, 116 pp., front. (map), plans. (Guy Le Strange, Nov. 1886).

2029
Muhammad Ghulām Imām Khān, called Hijr ibn Muhammad. Tārikh i Rashid al-din khāni. [A history of Hindustan and of the Deccan in particular; Urdu text]. Hyderabad, Mutba alia Khur-shidia, 1282 A.H. [1865 A.D.]. 765, 25 pp.

2030
Muhammad ibn al-Hasan, al-Nawāji. Halbat al-Kumait fi al-adab al-nawādir al-muta'alliqah bi'l-khamñyāt. [An anthology on wine and all that relates to it; edited by Nasr al-Hurini; Arabic text]. Cairo, 1299 A.H. [1881 A.D.]. 383, 3 pp.

2031 a d h
Muhammad ibn Jarir (Abū Ja'afar), al-Tabari. Chronique de Abou-Djafar-Mo'hammed-ben-Djarir-ben-Yezid Tabari; traduite sur la version persane d'Abou-Ali Mo'hammed Bel'ami . . . par Hermann Zotenberg. Paris, Imprimerie Impériale; Nogent-Le-Rotrou, Imprimerie de A. Gouverneur, 1867–74. 4 vols.

2032
Muhammad Sādiq, Bey. Mashál al-Mahmil. [Journal of the pilgrimage from Cairo to Mecca and back in the years 1297 and 1298 A.H.; Arabic text]. Cairo, Matbäh Wadi al-Nil, 1298 A.H. [1881 A.D.]. 60 pp., illus. [*see* p.].

2033 i
— (Nubthah) fi ishkshaf tariq al-ard al-Hijāzīyah. [The journal of a march from the port of al-Wijh i the Hijaz to Madina, and from thence back to the port of Yanbu'; reprinted from the journal 'al-Jandat al-'askariyah'; with a map of the route; Arabic text]. Cairo, Matba't 'Umūm Arkān Harb bi-Dīwān al-Jihādiyah, 1294 A.H. [1877 A.D.]. 28 pp., illus., map. [*see* p.].

2034 a d l
Murray, John, publisher. A handbook for India: being an account of the three Presidencies, and of thc overland route. Part II, Bombay. London, 1859. [iii], 241–591 pp., plan.

2035 a d h l
— A handbook for travellers in Syria and Palestine; new edition. London, 1868. 2 vols., maps (some in pockets).

2036 i
Mustafā ibn 'Abd Allāh, called Katib Chelebī, or Hājī Khalfah. Cronologia historica; scritta in lingua turca, persiana, & araba, da Hazi Halife' Mustafa', e tradotta nell'idioma italiano da Gio: Rinaldo Carli. In Venetia, Andrea Poletti, 1697. [viii], 206 pp. (J. Pincherle, Trieste, 10 May 1875).

2037 a d f
Nairne, A. K. The Konkan: an historical sketch. Bombay, Government Central Press, printer, 1875. [vi], 158 pp.

2038 a d
Nery Xavier, F. Instrucção do Exmo. Vice-Rei Marquez de Alorna ao seu successor o Exmo. Vice-Rei Marquez de Tavora; segunda edição. Nova-Goa, Imprensa Nacional, 1856. 3 pts. (in 1).

2039 d
Ockley, S. The history of the Saracens, comprising the lives of Mohammed and his successors; sixth edition. London, George Bell, 1878. xxviii, 512 pp., front. (Bohn's standard library).

2040 d i l
Oliphant, L. The land of Gilead, with excursions in the Lebanon. Edinburgh, London, William Blackwood, 1880. xxxvii, 538 pp., front., illus., plates, map.

2041 d
Oort, H. The worship of Baalim in Israel; translated . . . and enlarged . . . by John William Colenso. London, Longmans, Green, 1865. iv, 94 pp.

2042 d
Palestine Exploration Fund. Our work in Palestine: being an account of the different expeditions sent out to the Holy Land. London, Bentley, 1873. viii, 344 pp., front., illus., plates.

2043 a d h
Palgrave, W. G. Narrative of a year's journey through central and eastern Arabia (1862–63). London, Cambridge, Macmillan, 1865. 2 vols., map, plans.

2044 a d
Palmer, E. R. The desert of the Exodus journeys on foot in the wilderness of the Forty years' wanderings. Cambridge, Deighton, Bell, 1871. 2 vols., fronts., plates, maps.

2045 a d
Palmer, H. S. Sinai: from the fourth Egyptian dynasty to the present day. London,

Society for Promoting Christian Knowledge, [1878]. viii, 216 pp., front. (map), illus. At head of title: Ancient history from the monuments.

2046

Parkes, W. Euphrates valley railway: report on the ports of the Persian gulf, addressed to W. P. Andrew, Esq. London, Wm. H. Allen, 1872. 16 pp., maps.
Another copy. [*see* 258].

2047 d

Pierotti, E. Macpéla ou tombeau des Patriarches à Hébron. Lausanne, Howard et Delisle, 1869. 152 pp., plan. [*see* 258].

2048 a d

Playfair, R. L. A history of Arabia felix or Yemen. Bombay, printed for Government at the Education Society's Press, Byculla, 1859. (Selections from the records of the Bombay Government, N.S., Vol. 49).

2049 d

Porter, J. L. Five years in Damascus, with travels and researches in Palmyra, Lebanon, the giant cities of Bashan, and the Haurân; second edition, revised. London, John Murray, 1870. xvi, 339 pp., map.

2050 d h

— The giant cities of Bashan, and Syria's holy places. London, T. Nelson, 1869. 371 pp., fronts., plates.

2051

Price, W. H. Kurrachee harbor works. [Kurrachee, Commissioner's Press, 1874]. [ii], 29, [6] pp., map. [*see* pq.].

2052 a d

Rathbone, A. B. The true line of defence for India: a paper to be read before the East India Association on Wednesday, July 5, 1876. London, East India Association, 1876. 13 pp. [*see* p.].

2053

Rattray, H. Country life in Syria: passages of letters written from anti-Lebanon. London, Seeley, Jackson & Halliday, 1876. viii, 232 pp., front., plates.

2054 d h

Rawlinson, Sir H. England and Russia in the East: a series of papers on the political and geographical condition of central Asia. London, John Murray, 1875. xvi, 393 pp., map.

2055 d

Rivadeneyra, A. Viaje de Ceylan á Damasco, Golfo pérsico, Mesopotamia, ruinas de Babilonia, Níneve y Palmira. Madrid, Imprenta y Estereotipia de M. Rivadeneyra, 1871. xi, 399 pp.

2056 a d h

Robinson, E. and E. Smith. Biblical researches in Palestine, Mount Sinai and Arabia petraea: a journal of travels in the year 1838. London, John Murray, 1841. 3 vols. maps.

2057 k

Rowett, E. A journey up the Irrawaddy, from Rangoon to Bhamo. [privately printed], 1880. 38 pp., facsim.

2058 a d

Rueppell, E. Reisen in Nubien, Kordofan und dem peträischen Arabien, vorzüglich in geographisch-statistischer Hinsicht. Frankfurt am Main, Friedrich Wilmans, 1829. xxvi, 389 pp., front. (map), plates.

2059 a

Sádik, Isfaháni. The geographical works; translated by J. C. from original Persian MSS. in the collection of Sir William Ouseley, the editor [and] A critical essay on various manuscript works, Arabic and Persian . . . London, Oriental Translation Fund of Great Britain and Ireland, 1832. xiii, 152 pp.; xi, 71 pp.

2060

Sale, F. Lady. A journal of the disasters in Affghanistan, 1841–2; sixth thousand. London, John Murray, 1843. xvi, 451 pp., maps.

2061 d

Sayce, A. H. The Hittites: the story of a forgotten empire. London, The Religious Tract Society, 1888. 150 pp., illus. (By-paths of Bible knowledge, Vol. 12).

2062

Scherzer, K. von. La province de Smyrne, considérée au point de vue géographique, economique et intellectuel; traduit de l'Allemand par Ferdinand Silas. Vienne, Alfred Hölder, 1873. [viii], 258 pp., maps, tables. [*see* 230].

2063 b d l
Senior, N. W. A journal kept in Turkey and Greece in the autumn of 1857 and the beginning of 1858. London, Longman, Brown, [etc.], 1859. xiv, 372 pp., front. (map), plates, map.
2064 a d
Smith, G. Assyrian discoveries: an account of explorations and discoveries on the site of Niniveh, during 1873 and 1874; sixth edition. London, Sampson Low, [etc.], 1876. xvi, 461 pp., front. (map), plates, map.
2065 a d h
Smith, J. Y. Matheran hill: its people, plants, and animals. Edinburgh, Maclachlan & Stewart, 1871. [iii], 150 pp., front. (map).
2066
Socin, A. Bericht über neue Erscheinungen auf dem Gebiete der Palästinaliteratur 1880. [Leipzig, 1880]. 127-56 pp. Reprinted from *Zeitschrift des Palästina-Vereins,* Vol. 4. [*see* p.].
2067 a d
South Kensington Museum. Catalogue of the objects of Indian art exhibited, by H. H. Cole. London, Eyre & Spottiswoode, printer, 1874. x, 352 pp., front. (map), plates. At head of title: Science and Art Department of the Committee of Council on Education, South Kensington Museum.
2068 a d
Sprenger, A. Die alte Geographie Arabiens als Grundlage der Entwicklungs-geschichte des Semitismus. Bern, Huber, 1875. [i], 343 pp., maps (2 in pocket).
2069 d i
— Babylonien, das reichste Land in der Vorziet und das lohnendste Kolonisationsfeld für die Gegenwart. Heidelberg, Carl Winter's Universitätsbuchhandlung, 1886. [i], 171-296 pp., map. (Sammlung von Vorträgen, Vol. 15). [*see* p.].
2070
— The Ishmaelites, and the Arabic tribes who conquered their country. [n.p., n.d.]. 19 pp. [*see* 258].
2071 a d
[Steele, A.]. Summary of the law and custom of Hindoo castes within the Dekhun Provinces subject to the Presidence of Bombay, chiefly affecting civil suits . . . Bombay, Courier Press, printer, 1827. xvii, 331, 71, 15 pp.
2072 d i l
Thayer, A. W. The Hebrews and the Red Sea. Andover, Warren F. Draper, 1883. 140 pp., map.
2073 a d
Thomson, W. M. The land and the book: or, Biblical illustrations drawn from the manners and customs . . . of the Holy Land. London, T. Nelson, 1869. 718 pp., front., illus., map.
2074
Times, The. The Euphrates valley route to India; reprinted from "The Times" and "Financier". London, Henry S. King, 1871. 15 pp. [*see* 258].
2075 d h
Treitschke, H. von. La Turchia e le grandi potenze saggio; tradotto con licenza dell'autore da Silvio Sella. Roma, [etc.], Ermanno Loescher, 1876. 69 pp. [*see* p.].
2076 d
Tristram, H. B. The land of Israel: a journal of travels in Palestine; second edition. London, Society for Promoting Christian Knowledge, 1866. xx, 656 pp., front., plates, maps.
2077
Tschihatchef, P. di. Asia minore quale fu e quale è. [Rome, Tip. Succ. Le Minnier], 1873. 30 pp. Reprinted from *Nazione,* Vol. 15, Nos. 41-3, 45. [*see* 260].
2078
Turkey. Ottoman Empire. [Two Ottoman firmans giving English merchants trading privileges, and other consular business relating to Syria; Turkish text. Damascus, 1289 A.H. (1872 A.D.)]. 2 sheets (folded). [*see* p.].
2079 a d h
Upton, R. D. Gleanings from the desert of Arabia. London, C. Kegan Paul, 1881. viii, 399 pp.
2080 a d
Velde, C. W. M. van de. Narrative of a journey through Syria and Palestine in 1851 and 1852; translated under the author's superintendence. Edinburgh, London, William

Blackwood, 1854. 2 vols., fronts., plate, maps, table.

2081

— Notes on the map of the Holy Land; second edition. Gotha, Justus Perthes, 1865. [i], 48 pp.

2082 i

Verme, L. dal. Giappone e Siberia: note d'un viaggio nell'estremo oriente al seguito di S.A.R. Il Duca di Genova del Conte. Milano, Fratelli Treves, 1885. [vii], 487 pp., illus., maps. [*see* q.].

2083

Vincent, F. The land of the white elephant; notice par E.-A. Grattan. Anvers, Imprimerie Veuve de Backer, 1883. 16 pp. Reprinted from *Bulletin de la Société Royale de Géographie d'Anvers,* Vol. 8. [*see* p.].

2084 a d

Volney, C. F. Travels through Syria and Egypt, in the years 1783, 1784, and 1785; translated from the French; second edition. London, G. G. J. and J. Robinson, 1788. 2 vols., plates, maps.

2085 a d

Ward, W. A view of the history, literature, and mythology of the Hindoos; including a minute description of their manners and customs, and translations from their principal works; new edition. London, Kingsbury, Parburg & Allen, 1822. 3 vols.

2086

Watson, J. F. The Imperial Museum for India and the colonies. London, Wm. H. Allen, 1876. 62 pp., front. (map), illus.

2087 l

Watson, J. W. Statistical account of Júnágadh: being the Júnágadh contribution to the Káthiáwár portion of the Bombay gazetteer. Bombay, Bombay Gazette Steam Press, 1884. [vi], 160 pp., front. (port.), map, table.

2088 i

Watson, R. G. A history of Persia from the beginning of the nineteenth century to the year 1858. London, Smith, Elder, 1866. xii, 465 pp.

2089 d l

Watson & Co., publisher. The guide to Poona & Kirkee, containing list of residents and bungalows . . . 1875. Bombay & Poona, Watson, 1875. 59 pp., plates, map.

2090 a d

Wellsted, J. R. Travels in Arabia. London, John Murray, 1838. 2 vols., fronts., plates, maps.

2091 a d h k

Wilson, A. The abode of snow: observations on a journey from Chinese Tibet to the Indian Caucasus, through the upper valleys of the Himálaya. Edinburgh, London, William Blackwood, 1875. xxvi, 475 pp., front., map (in pocket).

2092 d h k

Wilson, Sir C. W. and Sir C. Warren. The recovery of Jerusalem: a narrative of exploration and discovery in the city and the Holy Land; with an introduction by Arthur Penrhyn Stanley; edited by Walter Morrison. London, Richard Bentley, 1871. xxvii, 554 pp., front., illus., plates, maps.

2093 a d

Wilson, H. H. Essays and lectures chiefly on the religion of the Hindus; collected and edited by Reinhold Rost. London, Trübner, 1862. 2 vols.

2094 a d

Wilson, J. Lecture on the religious excavations of western India. Bombay, Education Society's Press, Thacker, Vining, 1875. v, 74 pp. [*see* p.].

2095 a d

Wilson, W. R. Travels in the Holy Land, Egypt, etc.; fourth edition. London, Longman, Brown, [etc.], 1847. 2 vols., fronts., plates.

2096

Wood, W. M. 'Things of India' made plain: or, a journalist's retrospect. Part 1. London, Elliot Stock, 1884. 116 pp.

2097 a d

Wright, T., ed. Early travels in Palestine; with notes. London, Henry G. Bohn, 1848. xxxi, 517 pp., front. (Bohn's antiquarian library).

2098 a d

Zehme, A. Arabien und die Araber seit hundert Jahren: eine geographische und geschichtliche Skizze. Halle, Verlag der Buchhandlung des Waisenhauses, 1875. viii, 407 pp.

2099
Anonymous. Athens: its grandeur and decay. London, Religious Tract Society, [1848]. 192 pp.

2100 a d
— A character of France, to which is added, Gallus Castratus: or, an answer to a late slanderous pamphlet, called The character of England. London, Nath. Brooke, 1659. [viii], 45, [ii], 38 pp.

2101 l
— Cronaca veneta: sacra e profana . . . della citta' di Venezia. Venezia, Francesco Tosi, 1793. 2 vols., front., plates.

2102 d
— The Etruscans and their language. [London, 1875]. 405-37 pp. Reprinted from the *British quarterly review,* Vol. 62, Oct. 1875. [*see* 2388].

2103 d
— Hand-book to the Shetland islands. Kirkwall, William Peace, [1875?]. 114 pp., front., illus.

2104 i
— Lule-Lappmark: a sketch of Lapland travel. [London, C. Whiting, printer, 1869]. 83 pp. (E.B., 1 July 1869). [*see* p.].

2105 d h
— Outlines of the history of Ireland for schools and families, from the earliest period to the union in 1800 . . . second edition. Dublin, William Curry, Jun., 1847. viii, 382 pp., illus.

2106
— Pacte du Seigneur de Sarvantikar avec les Chevaliers de l'Ordre teutonique: document arménien de l'an 1271; traduction et notes. Venise, Imprimerie armenienne de Saint-Lazare, 1873. [11] pp., facsim. Reprinted from the *Journal polyhistor,* Vol. 31. [*see* p.].

2107 k l
— Rambles in Istria, Dalmatia and Montenegro, by R.H.R. London, Hurst & Blackett, 1875. xiii, 304 pp.

2108 d
— Scoperta della tomba del Duca Longobardo Gisulfo fatta in Cividale del Friuli, li 28 maggio 1874. [Cividale, Tip. Fanna], 1874. 8 pp. [*see* p.].
Another copy. [*see* 986].

2109 k
Nil.

2110 e l
— Sosivizka, the bandit of Dalmatia. [no imprint, 18—]. 560-76 pp. Reprinted from an unidentified journal, Vol. 32, No. 191. [*see* p.].

2111 l
Aedes pembrochianae: a new account and description of the statues, bustos, relievos, paintings, medals and other curiosities in Wilton-house . . . thirteenth edition. Salisbury, Salisbury Press, printers, 1798. xvi, 134, [14] pp., front. [*see* 261].

2112 a d
Ahmad ibn Muhammad ibn Ahmad ibn Yahya. The history of the Mohammedan dynasties in Spain; extracted from the Nafhu-t-tíb min ghosni-l-andalusi-r-rattíb wa tarikh lisánu-d-dín ibni-l-khattíb, by Ahmed ibn Mohammed al-Makkarí, a native of

Telemsán; translated from the copies in the library of the British Museum . . . by Pascual de Gayangos . . . London, printed for the Oriental Translation Fund of Great Britain & Ireland, 1840-3. 2 vols. [*see* q.].

2113 i

Aksákoff, I. Condensed speech of Mr. Ivan Aksákoff (Vice-President of the Slavonic Committee of Moscow), October 1876. [London, Robson, printers], 1876. (Olga Novikoff, née Kireéff. 12 pp. [*see* p.].

2114 i

Alishan, L. M. Physiographie de l'Arménie: discours prononcé le 12 aout 1861 à la distribution annuelle des prix au Collége arménien; seconde édition. Venise, Imprimerie arménienne de Saint-Lazare, 1870. 75 pp. [*see* p.].

2115 a d

Amati, A. and L. Tomaso. L'Istria sotto l'aspetto fisico, etnografico, amministrativo, storico e biografico studi. Milano, Francesco Vallardi, 1867. 24 pp., map. Reprinted from *L'Italia, sotto l'aspetto fisico, storico, artistico e statistico.* [*see* p.].

2116 a d

Anderson, J. Scotland in early Christian times. Edinburgh, David Douglas, 1881. 2 vols., fronts., illus., plates. (Rhind lectures in archaeology, 1879–80).

2117 l

Anquetil, L.-P. Histoire de France depuis les temps les plus reculés, jusqu'a la révolution de 1789; suivie de l'histoire de la République française, par M. de Norvins. Paris, Bureau des Publications illustrées, 1845–50. 5 vols., fronts., plates.

2118

Antas, M. d'. Les faux Don Sébastien: étude sur l'histoire de Portugal. Paris, Auguste Durand, 1866. [3], v, 477 pp.

2119 h

Ashridge House, Bucks. From Dan to Beersheba, through Ashridge Park. [London, 1863?]. [12] pp. Reprinted from an unidentified journal, pp. 348–59. [*see* p.].

2120 h

Austria. K. K. Kriegs-Archiv. Die Occupation Bosniens und der Hercegovina durch K. K. Truppen im Jahre 1878; nach authentischen Quellen dargestelt in der Abtheilung für Kriegsgeschichte des K. K. Kriegs-Archivs. Wien, Verlag des K. K. Generalstabes, 1879. 2 pts. Parts 1 and 3. [*see* p.].

2121 l

[Barratt, J.]. A description of the house and gardens at Stourhead. Bath, J. Barratt, 1822. 32 pp., front., plan. [*see* 261].

2122

Barrett, C. R. B. Caister castle. London, Lawrence & Bullen, 1893. 20 pp., front., illus. (Barrett's illustrated guides, No. 8). [*see* p.].

2123

— Colchester. London, Lawrence & Bullen, 1893. 42 pp., front., illus. (Barrett's illustrated guides, Eastern counties, No. 9). [*see* p.].

2124

— Round Ipswich. London, Lawrence & Bullen, 1893. 44 pp., front., illus. (Barrett's illustrated guides, No. 5). [*see* p.].

2125

— Round St. Osyth. London, Lawrence & Bullen, 1893. 41 pp., front., illus. (Barrett's illustrated guides, No. 3). [*see* p.].

2126

— St. Osyth. London, Lawrence & Bullen, 1893. 27 pp., front., illus. (Barrett's illustrated guides, No. 7). [*see* p.].

2127

— Yarmouth and Caister. London, Lawrence & Bullen, 1893. 50 pp., front., illus. (Barrett's illustrated guides, No. 6). [*see* p.].

2128

Barros e Cunha, J. G. de. To-day. London, W. H. Collingridge, 1868. 34 pp. At head of title: On the present situation, financial and political of the Kingdom of Portugal. [*see* 253].

2129 l

Bell, H. G. Life of Mary, Queen of Scots; third edition. London, Whittaker, 1840. xii, 163 pp. (Popular library of modern authors).

2130 d

[Benussi, B. Saggio d'una storia dell'Istria dai primi tempi sino all'epoca della dominazione romana]. Capodistria, Tipografia di Giuseppe Tondelli, 1872. 80 pp., table.

(Atti dell'I.R. Ginnasio superiore di Capodistria, 1871/2). [*see* p.].
d
Another copy. [*see* p.].
2131 a d
— compiler. Saggio d'una geografia dell'Istria; compilata ad uso della studiosa gioventù. Rovigno, Tpo-Litog. istr. di Antonio Coana, 1874. [68] pp., tables. [*see* p.].
2132 i
— — Manuale di geografia dell'Istria. Trieste, Stabilimento artistico, Tipografico G. Caprin, 1877. [iii], 132 pp., tables. (Trieste, 16 Feb. 1877).
Two other copies.
2133 i
Bianconi, G.-A. Il mare mediterraneo e l'epoca glaciale: memoria. Bologna, Tipi Gamberini e Parmeggiani, 1871. 53 pp. Reprinted from *Memorie dell'Accademia delle Scienze dell'Istituto di Bologna,* Serie 2, Vol. 10. [*see* pq.].
i
Another copy.
2134
Biasoletto, B. Escursioni botaniche sullo Schneeberg (Monte Nevoso) nella Carniola: discorso. Trieste, I. Papsch, 1846. 96 pp., front. (map). [*see* p.].
2135
Blake, C. C. Notes on human remains brought from Iceland by Captain Burton. [London, 1872]. 344-7 pp., plate. Reprinted from the *Journal of the Anthropological Institute,* Vol. 2. [*see* p.].
2136
— Sulphur in Iceland. London, E. & F. N. Spon, 1874. 51 pp., front., maps. [*see* p.].
2137 i j
Boemches, F. Der Bau des neuen Hafens in Triest. Wien, Lehmann & Wentzel, 1879. 25 pp., illus. Reprinted from the *Zeitschrift des öster. Ingenieur- und Architekten-Vereins,* 1879. (A. Bömches, Trieste, 2 April 1880). [*see* p.].
2138 a d
Boglić, G. Studi storici sull'isola di Lesina. Zara, Tip. di Giovanni Woditzka, 1873. 202 pp., tables. [*see* 2340].
2139 i
Bologna. Museo Civico. Nella solenne inaugurazione del Museo civico di Bologna fatta il 25 settembre 1881; discorso del Direttore generale Senatore Gozzadini. Bologna, Tipografia Fava e Garagnani, 1881. 21 pp., plates. [*see* p.].
2140 k
Bos, L. The antiquities of Greece; a new edition. London, T. Cadell & W. Davies, 1805. viii, 152 pp.
2141
Bowles, C. A short account of the Hundred of Penwith in the County of Cornwall . . . Shaftesbury, R. Hurd, 1805. iv, 46 pp. [*see* 261].
2142 i
Brandt, G. H. Royat (les bains) in Auvergne: its mineral waters and climate; second edition. London, H. K. Lewis, 1883. [v], 50 pp., front., plate, map.
2143
Britton, J. A historical account of Corsham house, in Wiltshire . . . London, Longman, [etc.], 1806. [iv], 108 pp., front. [*see* 261].
2144 f l
— and E. W. Brayley. Memoirs of the Tower of London: comprising historical and descriptive accounts of that national fortress and palace. London, Hurst, Chance, 1830. xvi, 374 pp., front., plates. imperfect.
2145
[Buchanan, G. and others]. Respublica, sive status regni Scotiae et Hiberniae, diversorum autorum. Lvgd. Bat., ex Officinâ Elzeveriana, 1627. 284 pp.
2146 a d
Buckle, H. T. History of civilization in England. London, John W. Parker, 1858-61. 2 vols.
2147 e
Burke, Sir J. B. Family romance: or, episodes in the domestic annals of the aristocracy; third edition. London, Hurst & Blackett, [1860]. vi, 341 pp., front.
2148 a e
Buzzi, I. L. preface. Dilucidazioni sul porto-canale Rieter. Trieste, Stab. Tip. Lit. di C. Coen, 1863. 23 pp. [*see* p.].

2149 a d
Calori, L. Della stirpe che ha popolata l'antica necropoli alla Certosa di Bologna e delle genti affini: discorso storico-antropologico. Bologna, Tipi Gamberini e Parmeggiani, 1873. 169 pp., plates, tables.
2150 a d i
— Intorno ai riti funebri degli Italiani antichi ed ai combusti del sepolcreto di Villanova e dell'antica necropoli alla Certosa di Bologna: dissertazione. Bologna, Tipi Gamberini e Parmeggiani, 1876. 58 pp., plate. [*see* pq.].
2151 a d
Campbell, Sir G. A handy book on the Eastern question: being a very recent view of Turkey; second edition. London, John Murray, 1877. xviii, 212 pp., map.
2152 i
Capellini, G. La formazione gessosa di castellina marittima e i suoi fossili: memoria. Bologna, Tipi Gamberini e Parmeggiani, 1874. 83 pp., plates. Reprinted from *Memorie dell'Accademia delle Scienze dell'Istituto di Bologna,* Series 3, Vol. 4. [*see* pq.].
2153 i
— L'uomo pliocenico in Toscana: memoria. Roma, Tipi del Salviucci, 1876. 17 pp., plates. Reprinted from *Atti della Reale Accademia dei Lincei,* Series 2, Vol. 3. [*see* pq.].
2154 d
Carrara, F. Topografia e scavi di Salona. Trieste, [Tipografia del Lloyd austriaco], 1850. viii, 172 pp., front. (map), map.
2155 d
Caruana, A. A. Report on the Phoenician and Roman antiquities in the group of the islands of Malta. Malta, Government Printing Office, 1822. viii, 168 pp., front., plates. [*see* q.].
2156 a d
Cassani, A. C. Saggio di proverbi triestini; raccolti ed illustrati da Angelo C-Cassani. Trieste, Tipografia di Colombo Coen, 1860. x, 110 pp. [*see* p.].
2157 a d h
Castellar, I. Nuevo compendio de la historia de España. Paris, T. H. Truchy, 1852. [5], 454 pp.
2158
Cattermole, R. The great Civil war of Charles I and the parliament. London, Longman, Orme, Brown, [etc.], 1841. ix, 288 pp., front., plates. (Cattermole's historical annual, Vol. 1).
2159 a
Ceresole, A. Montreux (lake of Geneva); from the French. Zürich, Orell Füssli; London, C. Smith, [1886]. 40 pp., front., illus., plates, map. [*see* p.].
2160 a d
— Vevey, its environs and climate. Zürich, Orell Füssli; London, C. Smith, [1884]. 40 pp., front., illus., plates, map. [*see* p.].
2161 a d
Chillon. Chillon ancient et moderne, description et histoire. Chillon, vente au Château de Chillon, [18—]. 31 pp. [*see* p.].
2162 i
Christoforo, N. La grandezza italiana studi confronti e desiderii. Torino, G. B. Paravia, 1864. xvi, 454 pp.
2163 a e
Clarke, H. The early history of the Mediterranean populations, &c., in their migrations and settlements. London, Trübner, 1882. 79 pp.
2164 e
— The Varini of Tacitus: or Warings, and their relations to English ethnology. [London], 1868. [18] pp. Reprinted from the *Transactions of the Ethnological Society,* N.S. Vol. 7. [*see* 256].
2165 a d
Clermont-Ferrand. Le vrai guide de Clermont-Ferrand et du Département du Puy-de-Dome; quatrième edition. Clermont-Ferrand, Duchier, [after 1865]. 179 pp. [*see* p.].
2166 h
Coglievina, F. Allerhöchste Reise: seiner kais. und kon. apostol. majestät Franz Josef I . . . durch Triest, Görz, nach Venedig, Istrien, Dalmatien und Fiume in . . . 1875. Wien, Selbstverlag des Verfassers, 1875. x, 447 pp.
2167 l
Collins, A. The peerage of England; containing a genealogical and historical account of

all the peers of that kingdom; fifth edition. London, W. Strahan, [etc.], 1779. 8 vols., fronts., plates.

2168
Collins, W. Rambles beyond railways: or, notes in Cornwall, taken a-foot; new edition. London, Richard Bentley, 1861. xv, 298 pp., front.

2169 a d
Conze, A., ed. Römanische Bildwerke einheimischen Fundorts in Osterreich. 1, Drei Sarkophage aus Salona. Wien, Karl Gerold, 1872. [ii], 20 pp., plates. Reprinted from *Denkschriften der philosophisch-historischen Classe der kaiserlichen Akademie der Wissenschaften,* Vol. 22. [*see* p.].

2170
Copenhagen. Musée des Antiquités du Nord. Guide illustré . . . par C. Engelhardt. Copenhague, [Gyldendal], 1876. [iv], 40 pp., illus.

2171 i
Corner, C. My visit to Styria. London, J. Burns, 1882. 32 pp. [*see* p.].

2172
Cornwall and Devonshire Mining Company. [Prospectus]. London, R. Clay, printers, 1825. 22 pp. [*see* 261].

2173 a d
Cowie, R. Shetland, descriptive and historical: being a graduation thesis on the inhabitants of the Shetland islands. Aberdeen, Lewis Smith, 1871. xvi, 310 pp., front. (map), plates.

2174 a d(?) i
Craveiro, T. A. Compendio da historia portugueza. Rio de Janeiro, Typ. de Ogier, 1833. vi, 246 pp., and Appendix, 48 pp. Appendix dated 1834.

2175 i
Crawfurd, O. Portugal old and new; new edition. London, Kegan Paul, Trench, [etc.], 1882. xi, 350 pp., front., map.

2176 e
Czoernig, C. Auf den Krn. (7095′ △, 2242 Meter). [Munich, 1872]. 317–18 pp. Reprinted from *Zeitschrift des deutschen und des oesterreichischen Alpenvereins,* Vol. 3. [*see* p.].

2177 d i
— Aus dem oberen Isonzo-Gebiete. [Munich], 1875. 243–56 pp., front. Reprinted from *Zeitschrift des deutschen und oesterreichischen Alpenvereins,* Vol. 6. [*see* p.].

2178 a d
— Aus Istrien: Rundtour um den Monte Maggiore . . . [Munich], 1873. 180–90 pp., illus. Reprinted from *Zeitschrift des deutschen und des oesterreichischen Alpenvereins,* Vol. 4. [*see* p.].

2179 i
— Die deutsche Sprachinsel Sauris in Friaul. [Munich, 1880]. 22 pp. Reprinted from *Zeitschrift des deutschen und oesterreichischen Alpenvereins,* Vol. 11. [*see* p.].

2180 h i j
— Die ethnologischen Verhältnisse des österreichischen Küstenlandes aus dem richtiggestellten Ergebnisse der Volkszählung vom 31. Dezember 1880. Triest, F. H. Schimpff, 1885. 37 pp., map. (Carl Czoernig, Klagenfurt, 21 Feb. 1885). [*see* p.].

2181
— Geschichte der Triester Staats-, Kirchen- und Gemeinde-Steuern: eine Darstellung des Ursprunges und der Entwicklung aller in dieser Stadt . . . Triest, F. H. Schimpff, 1872. [i], 109 pp., tables. [*see* p.].

2182
De Bosset, C. P. Proceedings in Parga and the Ionian islands. London, Longman, Hurst, Rees, [etc.]. 1819. xv, 198 pp., front. (map).

2183
Demidov, P. P., Prince di San Donato. The Jewish question in Russia; translated from the Russian . . . by J. Michell. London, Darling, 1884. x, 110 pp.

2184 a d
Denis, F. Une fête brésilienne célébrée à Rouen en 1550. Paris, J. Techener, 1850. 104 pp., plate. [*see* 251].

2185 a d h
Dennis, G. The cities and cemeteries of Etruria. London, John Murray, 1848. 2 vols., fronts., illus., plate, maps.

2186
Dick, W. R. A short sketch of the Beauchamp tower, Tower of London, and also a

guide . . . to the inscriptions and devices left on the walls thereof. [London, Grammer, printer, 187-]. 48 pp., front. [*see* p.].

2187

Dimitz, A. Geschichte Krains von der ältesten Zeit bis auf das Jahr 1813. Laibach, Ign. v. Kleinmayr & Fed. Bamberg, 1874-6. 4 pts. (in 2).

2188 a d

Draper, J. W. A history of the intellectual development of Europe; second edition. New York, Harper, 1864. [2], xii, 632 pp.

2189

Drinkwater, J. A history of the siege of Gibralter, 1779-1783, with a description and account of that garrison, from the earliest periods. London, John Murray, 1850. [iv], 172 pp., front. (map).

2190

Drumont, E. A. La France juive: essai d'histoire contemporaine; nouvelle édition. Vol. 2. Paris, C. Marpon & E. Flammarion, [1866]. [iii], 610 pp.

2191 a d g h

Du Chaillu, P. B. The Viking age: the early history, manners, and customs of the ancestors of the English-speaking nations. London, John Murray, 1889. 2 vols., fronts., illus., maps.

2192 a

Engel, W. H. Kypros: eine Monographie. Berlin, G. Reimer, 1841. 2 vols. condition

2193 a d

Evans, Sir A. J. The ancient bronze implements, weapons, and ornaments, of Great Britain and Ireland. London, Longmans, Green, 1881. xix, 509 pp., illus.

2194 a d

—The ancient stone implements, weapons, and ornaments, of Great Britain. London, Longmans, Green, 1872. xvi, 640 pp., illus., plates.

2195 i j

— Antiquarian researches in Illyricum. Westminster, [London], Society of Antiquaries, 1883-5. 4 pts. (in 2), illus., plates, maps. Reprinted from *The Archaeologia,* Vols. 48-9. (Gerald Massey, New Southgate, 16 July [1883?]). [*see* q.].

2196 a d h

— Through Bosnia and the Herzegovina on foot, during the insurrection, August and September 1875 with an historical review of Bosnia; second edition. London, Longmans, Green, 1877. civ, 445 pp., illus., plates, map.

2197 g h i

Faber, G. L. The fisheries of the Adriatic and the fish thereof: a report of the Austro-Hungarian sea fisheries. London, Bernard Quaritch, 1883. xxvi, 292 pp., front., plates, tables. Dedicated to Burton.

2198 a d f h

Farrer, J. Notice of runic inscriptions discovered during recent excavations in the Orkneys. [Edinburgh], printed for private circulation, 1862. 40 pp., front., plates.

2199 a

Ficker, A. Die Völkerstamme der österreichisch-ungarischen Monarchie, ihre Gebiete, Gränzen und Inseln: historisch, geographisch, statistisch. Wien. C. Ueberreuter'sche Buchdruckerei (M. Salzer); in Commission bei August Prandel, 1869. [ii], 96 pp., maps.

2200

Fiorelli, G. Guide de Pompéi; traduit de l'Italien par Eliezer Nicoletti. Napoli, [etc.], Enrico Detken, 1881. vii, 136 pp. [*see* p.].

2201 a d h

Fiume. Topografie von Fiume und Umgebung . . . Gedenkgabe für die XIV. Versammlung ungarischer Aerzte und Naturforscher; herausgegeben auf Kosten der Stadt Fiume. Wien, druck von Carl Gerold's Sohn, 1869. vi, 174 pp., illus., maps. [*see* 2199].

2202

Fligier, —. Ethnologische Entdeckungen im Rhodope-Gebirge. Wien, 1879. 34 pp. Reprinted from *Mittheilugen anthropologische Gesellschaft in Wien,* Vol. 9, pp. 165-96. [*see* p.].

2203

[Ford, R.]. Gatherings from Spain, by the author of the Handbook of Spain; chiefly selected from that work, with much new matter. London, John Murray, 1846. x, 342 pp.

2204 a d

Forel, F. A. Le lac léman: précis scientifique; deuxième édition. Bâle, Genève, Lyon,

H. Georg, 1886. [iv], 76 pp., tables.

2205 d

Formiggini, S. and others, eds. Three days at Trieste. [Trieste, Austrian Lloyd's Press, 1858]. vii, 127 pp., front., illus., plates. [*see* p.].

2206 d

Fortis, A. Saggio d'osservazioni sopra l'Isola di Cherso ed Osero. In Venezia, Gaspare Storti, 1771. [vi], 171 pp., front. (map), plates. [*see* 2207].

2207 a d

— Viaggio in Dalmazia. In Venetia, Alvise Milocco, 1774. 2 vols. (in 1). fronts., plates.

Iter Buda Hadrianopolim anno MXLIII exaratum ab Antonio Verantio [excudebat Venetiis, Aloysius Milocco, 1774]. xlviii pp., follows Vol. 1.

2208 i

Franceschi, C. de. L'Istria: note storiche. Parenzo, Tipografia di Gaetano Coana, 1879. 510 pp. (24 March 1880).

2209

[Frati, L.]. Tesoro monetale di bronzi primitivi scoperto in Bologna. [Bologna, Fava e Garagnani], 1877. 8 pp. Reprinted from *Gazetta dell'Emilia,* Vol. 47. [*see* p.].

2210 d

Freeman, E. A. The Ottoman power in Europe, its nature, its growth, and its decline. London, Macmillan, 1877. xxiii, 315 pp., front. (map), maps.

2211

— Sketches from the subject and neighbour lands of Venice. London, Macmillan, 1881. xix, 395 pp., front., plates.

2212 a d h

— The Turks in Europe. London, William Mullan, 1877. 61 pp. [*see* p.].

2213 a d

Gareis, A. Pola und seine nächste Umgebung. Triest, F. H. Schimpff, 1867. 94 pp., map. [*see* p.].

2214

Gastaldi, B. Deux mots sur la géologie des alpes cottiennes. Turin, Imprimerie Royale, 1872. 22 pp. Reprinted from *Comptes rendus de l'Académie des Sciences de Turin,* Vol. 7. [*see* p.].

2215 d

— Iconografia di alcuni oggetti di remota antichità: rinvenuti in Italia. Torino, Stamperia Reale, 1869. 50 pp., plates. Reprinted from *Memorie della Reale Accademia delle Scienze di Torino,* Series 2, Vol. 26. [*see* pq.].

2216

— Intorno ad alcuni fossili del Piemonte e della Toscana: breve note. Torino, Stamperia Reale, 1866. Reprinted from *Memorie della Reale Accademia delle Scienze di Torino,* Series 2, Vol. 24. [*see* pq.].

2217

— Lake habitations and pre-historic remains in the turbaries and marl-beds of northern and central Italy; translated by Charles Harcourt Chambers. London, Longman, Green for the Anthropological Society, 1865. xvi, 130 pp., illus., plates.

2218

— Raccolta di armi e strumenti di pietra delle adiacenze del Baltico. Torino, Stamperia Reale, 1870. 26 pp. Reprinted from *Atti della Reale Accademia delle Scienze di Torino,* Vol. 5. [*see* p.].

2219

— Scandagli dei laghi del Moncenisio, di Avigliana, di Trana e di Mergozzo. Torino, Stamperia Reale, 1868. 18 pp., plates. Reprinted from *Atti dell'Accademia delle Scienze di Torino,* Vol. 3. [*see* p.].

2220 d

— Sulla cossaite varietà sodica di Onkosina: breve nota. Torino, G. B. Paravia, 1875. 15 pp., plate. Reprinted from *Atti della Reale Accademia delle Scienze di Torino,* Vol. 10. [*see* p.].

2221

— Sur les glaciers pliocéniques de Mr E. Desor: note. Turin, J. B. Paravia, 1875. 20 pp. Reprinted from *Atti della Reale Accademia delle Scienze di Torino,* Vol. 10. [*see* p.].

2222

Geneva. Association des Intérêts du Commerce et de l'Industrie. Guide illustré de Genève. Genève, Imprimerie Chapalay et Mottier, [1886?]. 155 pp., front., illus., plates, map.

2223 d

— — — revised edition. Genève, Imprimerie de la "Tribune de Genève", 1888. 168 pp., illus., plates, map.

2224 a d h

Geuter, K. ed. Ischl und seine Umgebungen . . . fünfte Auflage. Ischl, Gmunden, E. Mänhardt, 1881. 112 pp., xxxii pp., front., maps. [*see* p.].

2225 d i

Gilbard, G. J. A popular history of Gibralter: its institutions, and its neighbourhood on both sides of the straights, and a guide book to their principal places and objects of interest. Gibraltar, Garrison Library Printing Establishment, 1886. [12], viii, 264 pp., plate, map.

2226 a d

Gladstone, W. E. Bulgarian horrors and the question of the East. London, John Murray, 1876. 64 pp. [*see* p.].

a

Another copy (reprint in smaller format). [*see* p.].

2227 i

Goracuchi, J. A. de. Attraits de Trieste avec un aperçu historique. Trieste, Imprimerie du Lloyd Austro-Hongrois, 1883. v, 179 pp.

2228 d

[Gordon, C. G.]. Memorandum on the treaties of San Stephano and Berlin; printed for private circulation. London, Edward Stanford, 1880. 7 pp., maps. [*see* p.].

2229

Gotha. Alamach de Gotha: annuaire généalogique, diplomatique et statistique, 1889. Gotha, Justus Perthes, [1889]. xx, 1088 pp., front. (port.), plates.

2230 a d

Gozzadini, G. Di alcuni sepolcri della necropoli felsinea: ragguaglio. Bologna, Fava e Garagnani, 1868. 30 pp., illus. [*see* p.].

2231 a d i

— Di un antico sepolcro a ceretolo nel Bolognese: esposizione. Modena, G. T. Vincenzi e Nipoti, 1879. 33 pp., plate. [*see* p.].

2232 a d i

— Di due statuette etrusche e di una iscrizione etrusca: dissotterrate nell'Apennino bolognese: memoria. Roma, coi Tipi del Salviucci, 1883. 9 pp., plates. Reprinted from *Memorie della Classe di Scienze morali, Storiche e Filologiche,* Series 3, Vol. 11. [*see* pq.].

2233 i

— Di due stele etrusche: memoria. Roma, Tipografia della R. Accademia dei Lincei, 1885. 8 pp., plates. Reprinted from *Memorie della Classe di Scienze morali, storiche e filologiche,* Series 3, Vol. 12. [*see* pq.].

2234 a i

— Di un sepolcreto, di un frammento plastico, di un oggetto di bronzo, dell'epoca di Villanova: scoperti in Bologna. Bologna, coi Tipi Fava e Garagnani, 1887. 16 pp., plates. Reprinted from *Atti e memorie della R. Deputazione di storia patria per le Provincie di Romagna,* Series 3, Vol. 5. [*see* p.].

2235 a d

— Intorno ad alcuni sepolcri scavati nell'arsenale militare di Bologna: osservazioni. Bologna, Fava e Garagnani, 1875. 14 pp., plate. [*see* p.].

2236 a d i

— Intorno agli scavi archeologici fatti dal Sig. A. Arnoaldi veli presso Bologna: osservazioni. Bologna, Tipografia Fava e Garagnani, 1877. 96 pp., plates. [*see* q.].

a

Another copy.

2237

— La nécropole de Villanova: découverte et décrite. Bologne, Fava e Garagnani, 1870. 80 pp., illus. [*see* p.].

2238 a d i

— Note archeologiche per una guida dell'Apennino bolognese. Bologna, Tipografia Fava e Garagnani, 1881. 30 pp. [*see* p.].

2239 a d f i

— Note sur une cachette de fondeur ou fonderie à Bologne. Toulouse, Bonnal et Gibrac, printer, 1877. 12 pp. Reprinted from *Matériaux pour l'histoire primitive et naturelle de l'homme,* Series 2, Vol. 8, pp. 249–58. [*see* p.].

2240 a i
— Nuovi scavi nel fondo S. Polo presso Bologna. Roma, coi Tipi del Salviucci, 1884. 19 pp. Reprinted from *Atti: notizie degli scavi di antichita, Accademia dei Lincei,* 1884. [*see* pq.].
2241 a i
— Nuovi scavi del podere S. Polo presso Bologna. Roma, coi Tipi del Salviucci, 1884. 17 pp. [*see* p.].
2242 a d f
— Renseignements sur une ancienne nécropole à Marzabotto, près de Bologne. Bologne, Fava e Garagnani, 1871. 19 pp., illus. [*see* p.].
2243 a j
— Il sepolcreto di crespellano nel Bolognese. Bologna, Tipografia Fava e Garagnani, 1881. 12 pp., plate. (G. Gozzadini, Bonzano, 24 Aug. 1881). [*see* p.].
2244 a j
Gregor, W. An echo of the olden time from the north of Scotland. Edinburgh, Glasgow, John Menzies; Peterhead, David Scott, 1874. 168 pp. (Walter Gregor, Fraserburgh, 27 Nov. 1883).
2245 h i j
Gregorutti, C. Le antiche lapidi di Aquileja. Trieste, Julius Dase, 1877. xviii, 284 pp. (V. Kaltenegger, Laibach, 17 Oct. 1877).
2246 i
— Esemplare di una decorazione militare fomana della categoria delle falere. Trieste, [1877]. 15 pp., illus., plate. Reprinted from *Archeografo triestino,* N.S., Vol. 5, No. 2. [*see* p.].
2247 a
Gualandi, M. Guida di Bologna e suoi dintorni; quarta edizione. Bologna, Nicola Zanichelli, 1875. viii, 195 pp., plates, map.
2248 a
Gudjónsson, P. Íslenzk sálmasaungs- og messubók med nótum . . . gefin út af hinu Íslenzka bókmentafèlagi. Kaupmannahöfn, prentud Hjá Thiele, 1861. [iv], 180 pp., music. [*see* p.].
2249 a d h
Haas'sche Erben, M. ed. Steyr in Ober-Oesterreich und seine nächsten Umgebungen. Steyr, M. Haas'sche Erben, printer, 1877. 130 pp., plates. [*see* p.].
2250 d h
Hahn, J. G. von. Albanesische Studien. Jena, Friedrich Mauke, 1854. 3 pts. (in 1), illus., plates, maps.

2251 k
Heiss von Kogenheim, J. Histoire de l'Empire contenant son origine, ses progrès ses révolutions . . . Amsterdam, Wetsteins & Smith, 1733. 8 vols.
2252 a d h
Hellbach, R. Der kundige Begleiter auf den Vergnügungszügen zwischen Wien und Triest. Wien, Verlag von J. Dirnböck's Buchhandlung, [1878]. viii, 174 pp., map.
2253 a d
[Hellen, R.]. Letters from an Armenian in Ireland, to his friends at Trebisond, &c.; translated in the year 1756. London, printed for W. Owen, 1757. 250 pp.
2254 a d f
Henderson, E. Iceland: or the journal of a residence in that island, during the years 1814 and 1815; containing observations on the natural phenomena, history, literature, and antiquities of the island; and the religion, character, manners, and customs of its inhabitants; with an introduction and appendix. Edinburgh, printed for Oliphant, Waugh & Innes; London, Hamilton, J. Hatchard & L. B. Seeley, 1818. 2 vols., plates, map.
2255
Herzog, B., compiler. Die Eisenbahn-Verbindungen von Marienbad nach dem In- und Auslande. Marienbad, Josef Gschihay, [18—]. 31 pp., map. [*see* p.].
2256 l
Hjaltalín, J. A. The thousandth anniversary of the Norwegian settlement in Iceland. Reykjavík, Einar Iórdarson, printer, 1874. 34 pp. [*see* p.].
2257 a d
Holmes, G. Sketches of some of the southern counties of Ireland, collected during a tour in the autumn, 1797 in a series of letters. London, Longman & Rees, [etc.], 1801. vii, 211 pp., front., plates.

2258
Impastari, M. A. Muggia e il suo vallone. Trieste, Stabilimento Artistico Tipografico G. Caprin, 1896. 51 pp., plates.
2259
Irving, W. A chronicle of the conquest of Granada, from the mss. of Fray Antonio Agapida . . . to which is added Legends of the conquest of Spain by Washington Irving. London, George Bell, 1877. 2 vols. (in 1). (Bohn's standard library).
2260 d
Joanne, A. Géographie du Département de la Loire. Paris, Hachette, 1874. x, 56 pp., illus., maps.
2261 a d h
Jung, G., ed. Guide to Salzburg and environs; reserved for visitors at the Hôtel de l'Europe; third edition. Salzburg, [1880]. [2], iv, 88 pp., front., map, plan. [*see* p.].
2262
[Kandler, P.]. L'emporio e il Portofranco. [Trieste, Tipografia del Lloyd Austriaco, after 1864]. 299 pp.
2263 h
— Il foro di marte in Parenzo. Trieste, Tip. Coen, 1858. 4 pp., illus., plates. From *Atti del Conservatore,* No. 1, febbraro 1858. [*see* 2264].
2264 h l
— Indicazioni per riconoscere, le cose storiche del Litorale. Trieste, Tipografia del Lloyd Austriaco, 1855. xii, 296 pp., front., plans. [for offprints bound in *see* 336, 2263].
2265 a d
— Notize storiche di Montona; con appendice. Trieste, Tipografia del Lloyd Austro-Ungarico, 1875. [iii], 292 pp., table.
2266
— Per nozze Guastalla Coen-Ara: discorso sull'Istro adriaco. Trieste, Tipografia del Lloyd Austriaco, 1867. 16 pp., plate. [*see* p.].
2267 a d
— Per nozze Guastalla-Levi: discorso sul Timavo. Trieste, Tipografia Lloyd Austriaco, 1864. 41 pp., maps. [*see* p.].
2268
— Per occasione di via Ferrata proposta fra il Dravo di Carintia e Trieste pel varco piciano (prediel): discorso sulla Giulia e sulle strade antiche che la attraversavano. Trieste, Tipografia del Lloyd Austriaco, 1867. [iv], 24 pp. [*see* p.].
2269
Kent, T. One month in Iceland: a diary. Southampton, Gutch & Cox, printer, 1875. 79 pp. [*see* p.].
2270
Knittl, M. Cilli. Cilli, Fritz Rasch, 1890. 192 pp., front., plates. [*see* p.].
2271 h
Kohl, J. G. Ireland, Dublin, the Shannon, Limerick . . . London, Chapman & Hall, 1843. [ii], 248 pp. Bound in: The ceremonies of the Holy Week in the Papal Chapel at the Vatican, Rome, Monaldini, 1867, 36 pp.
2272
Kukuljević Sakcinski, I. Njeke gradine i gradovi u Kraljevini Hrvatskoj. U Zagrebu, Stamparna Dragutina Albrechta, 1869–70. 3 pts. [*see* pq.].
2273 a d
Kunz, C. and C. Gregorutti. Il Museo civico di Antichità di Trieste: informazione di Carlo Kunz con note illustrative del lapidario triestino del Dre Carlo Gregorutti. Trieste, Tip e Calc. di G. Balestra & C., 1879. 103 pp., plates.
2274 a d
Landnámabók. Islands landnámabók; hoc est: liber originum islandiae; versione latina . . . ex manuscriptis Legati Magnaeani. [The recension by Björn Jónsson of Skarosá; edited and translated by Jón Finnsson, Bishop of Skálholt, with index of rare and poetic words by Jón Olefsson]. Havniae, Typis Augusti Frederici Steinii, 1774. [xxii], 518 pp.
2275
Lane-Clarke, L. Folk-lore of Guernsey and Sark. Guernsey, E. Le Lievre, printer, 1880. vii, 144 pp.
2276 d
Lane Fox, A. H. Roovesmore fort, and stone inscribed with oghams, in the parish of Aglish, County Cork. [London, 1867]. 17 pp., front., plates. Reprinted from the

Archaeological journal, Vol. 24, pp. 123-39. [*see* 241].

2277 a d
Lanza, F. Antiche lapidi salonitane; inedite illustrate . . . seconda edizione. Zara, Tipografia Battara, 1850. [vii], 168 pp.

2278 a d
— Dell'antico palazzo di Diocleziano in Spalato; illustrazione con dodici tavole tratte dall'originale. Trieste, Tip. del Lloyd Austriaco, 1855. 29 pp., plates. [*see* pq.].
Another copy.

2279 a
Lartet, E. and H. Christy. Cavernes du Périgord: objets gravés et sculptés des temps pré-historiques dans l'Europe occidentale. Paris, aux Bureaux de la Revue archéologique, 1864. 37 pp. Reprinted from *Revue archéologique,* N.S. Vol. 9. [*see* 222].

2280 j
Lavardin, J. de. The historie of George Castriot, surnamed Scanderberg, king of Albanie, containing his famous actes, his noble deedes of armes . . . London, imprinted for William Ponsonby, 1596. [x], 498, [17] pp. (Bernard Quaritch, 4 June 1875).

2281 a d i
Lemos Pereira de Lacerda, J. A., Visconde de Juromenha. Cintra pinturesca, ou memoria descriptiva da villa de Cintra, Collares, e seus arredores. Lisboa, Typographia da Sociedade propagandora dos Conhecimentos Uteis, 1838. 232 pp. Upper wrapper dated 1839.

2282 d i
Littrow, E. de. Fiume considerata dal lato marittimo. Fiume, Stabilimento Tipolitografico di Emidio Mohovich, 1870. 44 pp., front., maps, tables. [*see* p.].
a i
Another copy.

2283 a i
— Fiume und seine Umgebungen. Fiume, Tipo-lithographische Anstalt des Edmidio Mohovich, 1884. [iii], 124 pp., plates, map, plan.

2284
Ljubić, S. Dispacci di Luca de Tollentis vescovo di Sebenico e di Lionello Cheregato vescovo di Traù, Nunzi Apostolici in Borgogna e nelle Fiandre 1472-1488. Zagrabria, Tipografia sociale, 1876. 64 pp. [*see* p.].

2285 a d
— Faria città vecchia e non Lesina. Zagrabria, Stamperia di Carlo Albrecht, 1873. 68 pp.

2286
— Književna obznana. [Zagreb?, 1874]. 17 pp. [*see* p.].

2287
— O Markantunu Dominisu rabljaninu: navlastito po izvorih mletačkoga arkiva i knjižnice arsenala parizkoga: historičko-kritičko iztraživanje. U Zagrebu, Stamparija Dragutina Albrechta, 1870. [ii], 159 pp. Reprinted from *Rada,* Vol. 10. [*see* p.].

2288
— Popis predmeta iz predhistoričke dobe u Nar. Zem. Muzeju u Zagrebu. U Zagrebu, C. Albrecht, 1876. [i], 56 pp., plates. [*see* p.].

2289
— Prilozi za životopis Markantuna de Dominisa Rabljanina, spljetskoga nadbiskupa. U Zagrebu, Dragutina Albrechta, 1870. [i], 260 pp. (Starina, Vol. 2).

2290 k
Longman, W. The history of the life and times of Edward the Third. London, Longmans, Green, 1869. 2 vols., front., plates, maps.

2291 a d h
— Suggestions for the exploration of Iceland: an address delivered to the members of the Alpine Club on April 4, 1861; second edition. London, Longman, Green, [etc.], 1861. 44 pp., map (front.). [*see* p.].

2292 l
Lowe, L. The bastilles of England: or, the lunacy laws at work. Vol. 1. London, Crookenden, 1883. [viii], 153 pp., front. (port.).

2293 a
Lucca, S. Zur Orientirung in Marienbad: ein Rathgeber und Wegweiser für Curgäste. Marienbad, Druck und Verlag von Josef Gschihay, 1883. 163 pp., maps.

2294 a
Luciani, T. Albona studii: storico-etnografici. Venezia, Tipografia dell'Istituto Coletti, 1879. [iii], 32 pp. [*see* p.].

a
Two other copies.
2295 a d
— L'Istria. [Venezia?, after 1862]. 8–103 pp. Reprinted from an unidentified journal. [*see* p.].
2296 a e
— Mattia Flacio, Istriano di Albona: notizie e documenti. Pola, Tipografia G. Seraschin, 1869. 24 pp. [*see* p.].
2297 d h i
MacColl, M. Three years of the Eastern question; third edition. London, Chatto & Windus, 1878. [vii], 302 pp.
2298 d h
MacGahan, J. A. The Turkish atrocities in Bulgaria: letters of the special commissioner of the "Daily News"; with an introduction and Mr. Schuyler's preliminary report. London, Bradbury, Agnew, 1876. 94 pp. [*see* p.].
2299 a d h
[Mackay, G. E.]. A week in Venice: a complete guide-book to the city and its environs; second edition. Venice, [etc.], Coen's New Library, 1870. 176 pp., front., illus., maps. [*see* p.].
2300 a d h
Mallet, P. H. Northern antiquities . . . of the ancient Scandinavians; translated from the French . . . by Bishop Percy; new edition . . . by I. A. Blackwell. London, Henry G. Bohn, 1859. [vi], 578 pp., front. (Bohn's antiquarian library).
2301 a d f l
Marazzi, A. La pesca lungo le coste austro-ungarische e la flottiglia peschereccia italiana: memoria. Roma, Tipografia di E. Sinimberghi, 1873. 69 pp., tables. Reprinted from *Bollettino consolare.* [*see* p.].
2302 d
[Marchesetti, C.]. Alcune mostruosità della flora illirica. [Trieste, 1878]. 4 pp., plate. Reprinted from *Bollettino della Società adriatica di Scienze naturali in Trieste.* Vol. 3. [*see* p.].
2303 d
— Di alcune nuove località del Proteus anguinus, Laur – Flora dell'isola S. Catterina presso Rovigno – Della presenza di piante alpine nelle paludi del Friuli. [Trieste, 1875]. 13 pp., diagr. Reprinted from *Bollettino della Società adriatica di Scienze naturali in Trieste,* Vol. 1. [*see* p.].
2304 a d
— Una escursione alle alpi giulie. [Trieste, 1880?]. 46 pp. [*see* p.].

2305 a d i
— Florula del campo marzio. [Trieste], 1882. 14 pp. Reprinted from *Bollettino della Società adriatica di Scienze naturali,* Vol. 7, No. 1. [*see* p.].
2306 d
— Una gita al Gran Sasso d'Italia. [Trieste, 1875]. 11 pp. Reprinted from *Bollettino della Società adriatica di Scienze naturali in Trieste,* Vol. 1 [*see* p.].
2307
— La necropoli di S. Lucia presso Tolmino . . . scavi del 1884. Trieste, 1886. [i], 73 pp., plates, tables. Reprinted from *Bollettino della Società adriatica di Scienze naturali in Trieste,* Vol. 9, No. 2. [*see* p.].
2308 d i
— Particolarità della flora d'Isola. [Trieste, 1879]. 6 pp. Reprinted from *Bollettino della Società adriatica di Scienze naturali in Trieste,* Vol. 4. [*see* p.].
2309 d i
— Una passeggiata alle alpi carniche. [Trieste, 1879]. 23 pp. Reprinted from *Bollettino della Società adriatica di Scienze naturali in Trieste,* Vol. 4. [*see* p.].
2310 d i
— Del sito dell'antico castello pucino e del vino che VI Cresceva. [Trieste, 1877]. 20 pp. Reprinted from *Archeografo triestino,* N.S., Vol. 5, No. 4. [*see* p.].
d h
Another copy.
2311 d i
— Sugli oggetti preistorici scoperti recentemente a S. Daniele del Carso. [Trieste, 1879]. 13 pp., plates. Reprinted from *Bollettino della Società adriatica di Scienze naturali in Trieste,* Vol. 4. [*see* p.].

2312 d
Marenzi, Graf F. von. Der Karst: ein geologisches Fragment im Geiste der Einsturztheorie; zweiter Manuscript-Abdruck. Triest, Buchdruckerei des Oesterreichischen Lloyd, 1865. 23 pp., map. [*see* p.].

2313 d
Marieni, G. Portolano del mare adriatico; compilato sotto la direzione dell'Istituto geografico militare. Milano, Imperiale Regio Stamperia, 1830. xii, 599 pp. [*see* q.].

2314
Matthews, H. The diary of an invalid: being the journal of a tour in pursuit of health in Portugal, Italy, Switzerland and France in the years 1817, 1818 and 1819; second edition. London, John Murray, 1820. xv, 515 pp.

2315 a d j
Michel, F. Le pays basque: sa population, sa langue, ses moeurs, sa littérature et sa musique. Paris, Firmin Didot, [etc.], 1857. [iii], 547 pp. (J. F. Blumhardt, Harrogate, 25 Dec. 1886; 2 Jan. 1887).

2316 k
[Millot, C. F. X., Abbé]. Abrégé de l'histoire romaine. Paris, chez Nyon l'aîné & Fils, 1789. viii, 192 pp., front., plates.

2317 d h
Mills, J. The British Jews. London, Houlston & Stoneman, 1853. xii, 415 pp., front., illus.

2318 a d
Moore, T. The history of Ireland, from the earliest kings of that realm, down to its last chief. London, Longman, Brown, [etc.], [1835-46]. 4 vols. Vols. 2-3 are of the 'new edition'.

2319 k
[—]. Memoirs of Captain Rock, the celebrated Irish chieftain, with some account of his ancestors; written by himself. Paris, A. & W. Galignani, 1824. xv, 311 pp.

2320
Morchio, G. F. M. and G. Morchio, publishers. Tavole sinottiche nummografiche della Republica di Venezia, rappresentanti la classazione e rarità delle monete, ducali e dei possedoenti. Venezia, Litografia G. Corradini, 1878. 11 pp. [*see* pq.].

2321 i
Muñoz y Gaviria, J. Historia del Alzamiento de los Moriscos: su espulsion de España. Madrid, Tipografico de Mellado, 1861. vii, 195 pp. (Santa Isabel, 8 de setiembre de 1862).

2322 a d
Murray, John, publisher. A handbook for travellers in central Italy. Part II: Rome and its environs; fourth edition. London, John Murray, 1856. xxiv, 356 pp., maps.

2323
— A handbook for travellers in central Italy; seventh edition. London, John Murray, 1867. xi, 480 pp., map.

2324 d
— A handbook for travellers in Denmark, Norway, and Sweden; third edition, revised. London, John Murray, 1871. 3 pts. (in 1), maps.

2325 a d
— Handbook for travellers in Ireland. London, John Murray, 1864. lxvi, 354 pp., maps (1 in pocket).

2326 a d
— Handbook for travellers in northern Italy; sixth edition. London, John Murray, 1856. xxix, 414 pp., maps, (wanting map in pocket).

2327 d
— Handbook for travellers in northern Italy; eleventh edition. London, John Murray, 1869. xxxvi, 620 pp., maps (1 in pocket).

2328 a d
— A handbook for travellers in Portugal; third edition. London, John Murray, 1864. lxiv, 216 pp., maps (1 in pocket).

2329 a d
— A handbook for travellers in Sicily. London, John Murray, 1864. lvi, 524 pp., maps.

2330 d l
— Handbook for travellers in southern Germany; eleventh edition, revised. London, John Murray, 1871. xii, 635 pp., maps.

2331
— A handbook for travellers in southern Italy; sixth edition. London, John Murray,

1868. xlv, 470 pp., maps (1 in pocket).

2332 a d

— A handbook for travellers in Spain, by Richard Ford; third edition. London, John Murray, 1855. 2 vols., maps (1 in pocket).

2333 l

— A handbook for travellers in Switzerland, and the Alps of Savoy and Piedmont; seventh edition. London, John Murray, 1856. lxxi, 430 pp., maps.

2334 l

— A handbook for travellers in Wiltshire, Dorsetshire, and Somersetshire; new edition. London, John Murray, 1859. iv, 272 pp., map (in pocket).

2335 l

— A handbook of Rome and its environs; eleventh edition. London, John Murray, 1872. lviii, 498 pp., maps.

2336 d i j

— Handbook to the Mediterranean . . . by . . . R. L. Playfair. London, John Murray, 1881. 2 pts. (in 1), fronts. (maps), maps. (to Lady Burton; John Murray, London, 16 Nov. 1889).

d

Another copy.

2337

— London as it is, by Peter Cunningham. London, John Murray, [1863]. lvi, 312 pp., maps. Title on spine: Hand-book of modern London.

2338 d

Nadeau, L. Vichy: son passé, son présent et son avenir. Vichy, Imprimerie A. Wallon, [1869?]. 53 pp., illus. [*see* p.].

2339

Niccolini, A. Quadro in musaico: scoperto in Pompei a di 24 ottobre 1831. Napoli, dalla Stamperia Reale, 1832. [ii], 91, xxiv pp., plates.

2340 a d

Nicolich, M. Storia documentata dei Lussini. Rovigno, Tipo-litog. istriana di Antonio Coana, 1871. 280 pp.

2341

Nicolucci, G. Strumenti in pietra delle Provincie calabresi: memoria. Napoli, Tipografia dell'Accademia Reale delle Scienze, 1879. [iii], 19 pp., plates. Reprinted from *Atti della Reale Accademia delle Scienze fisiche e matematiche di Napoli,* Vol. 8. [*see* pq.].

2342 d

O'Kelly, P. History of Ireland, since the expulsion of James II, by his son-in-law William III, Prince of Orange. Dublin, printed for the author by Goodwin, Son & Nethercott, 1855. 384 pp.

2343

Park Jurjavés. [Vienna, Carl Gerold, *c.* 1853]. 10 pp., front., plates, map. [*see* q.].

2344

Peacock, E., ed. The Army lists of the Roundheads and Cavaliers, containing the names of the officers in the Royal and Parliamentary armies of 1642; second edition, revised, corrected, and enlarged. London, Chatto & Windus, 1874. xii, 128 pp.

2345

Pecorini-Manzoni, C. Storia della 15A Divisione Türr nella campagna del 1860 in Sicilia e Napoli. Firenze, Tipografia della Gazzetta d'Italia, 1876. xii, 531 pp., maps.

2346 d

Peirce, B. M., compiler. A report on the resources of Iceland and Greenland. Washington, Government Printing Office, 1868. [i], 72 pp., maps, tables. At head of title: U.S. State Department. [*see* p.].

2347 l

Petit, P. De Amazonibus dissertatio quâ an verè extiterint, necne, variis ultro citroque conjecturis & argumentis disputatur. Amstelodami, Johannem Walters & Ysbrandum Haring, 1687. [xii], 396 pp., front., illus., map.

2348 a d

— Traité historique sur les Amazones: où l'on trouve tout ce que les auteurs . . . ont écrit pour ou contre ces héroines . . . par Pierre Petit. Leide, J. A. Langerak, 1718. 2 vols. (in 1), fronts., illus., plate.

2349 a e

Pètursson, P. Historia ecclesiastica Islandiae, ab anno 1740, ad annum 1840. Havniae, Typis excudebat Bianco Luno, 1841. [vii], 508 pp.

2350 a d i
Pichler, R. Il castello di Duino: memorie. Trento, Stabilimento Tipografico di Giovanni Seiser, 1882. viii, 474 pp., plate, tables.
2351
Pitts, J. L. Witchcraft and devil lore in the Channel Islands. Guernsey, Guille-Allès Library; Thomas M. Bichard, printer, 1886. [viii], 40 pp. [*see* p.].
2352 d
Planche, J. R. Descent of the Danube, from Ratisbon to Vienna, during the autumn of 1827, with anecdotes and recollections . . . of the towns, castles, monasteries, &c., . . . London, James Duncan, 1828. xv, 320 pp., front., map.
2353 a d
Pola. Cura del Municipio. Notizie storiche di Pola; edite per. Cura del Municipio e dedicate agli onorevoli membri della Società agraria istriana radunati al IX Congresso generale nella Città di Pola. Parenzo, Tipografia di Gaetano Coana, 1876. [ii], 437 pp., plates, map.
2354 a d
Provost & Co., publisher. Provost's guide to Gibraltar, containing the history, geography, chief attractions . . . of the Rock. London, Provost; Gibraltar, Chronicle Office, [1869]. 47 pp., front. (map). [*see* p.].
2355 d h
Rabl, J. Curort Abbazia. Wien, Verlag der K.K. Priv. Südbahn-Gesellschaft, 1887. 22 pp., front., plans. At head of title: Prospect. [*see* p.].
2356 i
Rački, F. O Dalmatinskih i Ilirskih novcih najstarije dobe. U Zagrebu, Stamparija Dragutina Albrechta, 1871. [i], 43 pp., plates. [*see* p.].
2357 d h
Rieger, G. Panorama della cost e delle isole di Dalmazia, nei viaggi dei Piroscafi del Lloyd Austriaco; disegnato per ordine dello stabilimento suddetto. Trieste, Sezione Letterario-Artistica del Lloyd Austriaco, 1863. 40 folded leaves. (annotations: Sebenico section; insertions). [*see* p.].
2358 d i
Rodwell, G. F. Etna: a history of the mountain and of its eruptions. London, C. Kegan Paul, 1878. xi, 146 pp., front., plates, maps. (30 Jan. 1881).
2359 a d
Rohitsch-Sauerbrunn dans la Basse-Styrie. [Graz, 1885]. 63 pp., illus., plate, map. [*see* p.].
2360
Rory-O'-the-Hills, pseud. Letter to a grand old man, and certain cabinet ministers; second edition. London, Tinsley, 1882. 64 pp. At head of title: Practical politics and moonlight politics. [*see* p.].
2361 a d
Royal Irish Academy. Museum. A descriptive catalogue of the antiquities of gold, . . . by W. R. Wilde. Dublin, Hodges, Smith; London, Williams & Norgate, 1862. [iii]. 100 pp., illus.
2362 d l
[Royer, C.]. Le lac de Paris: essai de géographie quarternaire. [Versailles], 1877. 46 pp. Reprinted from *Philosophie positive,* mars-avril 1877. [*see* p.].
2363 d
Russell, J. 1st Earl. An essay on the history of the English government and constitution from the reign of Henry VII to the present time; new edition. London, Longman, Green, [etc.], 1865. cviii, 378 pp.
2364
Rutter, J. A brief sketch of the history of Cranborn Chace, and of the dispute concerning its boundaries. Shaftesbury, J. Rutter, 1818. 16 pp. [*see* 261].
2365 l
— An historical and descriptive sketch of Wardour castle and demesne, Wilts, the seat of Everard Arundell, Lord Arundell of Wardour. Shaftesbury, J. Rutter, 1822. [v], 53 pp., front. [*see* 261].
2366
— A new descriptive guide to Fonthill abbey and demesne, for 1823 . . . Shaftesbury, J. Rutter, 1823. viii, 98 pp., front., plate, plan. [*see* 261].
2367
Rzehak, A. Neu Entdeckte prähistorische Begräbnisstätten bei Mönitz in Mähren. Wien, 1879. 15 pp., plates, diagr. Reprinted from *Mittheilungen der anthropologischen*

Gesellschaft in Wien, Vol. 9. [*see* p.].

2368 d l

Ržiha, F. Die Bedeutung des Hafens von Triest für Österreich. Prag, 1873. 39 pp., tables. Reprinted from *Technischen Blättern. Vierteljahrschrift des deutschen polytechnischen Vereines in Böhmen,* Jahrgang 5, Heft 3. [*see* p.].

2369 l

Schaffarik, P. J. Über die Abkunft der Slawen nach Lorenz Surowiecki. Ofen, mit Kon. Ung. Universitäts-Schriften, 1828. 212 pp.

2370 a d

Schio, G. da. Le antiche iscrizioni che furono trovate in Vicenza. Bassano, Tipografia Baseggio, 1850. 128 pp., plates.

2371 a d i

— Di due astrolabj in caratteri cufici occidentali trovati in Valdagno: comunicazione seconda. [Venezia, 1880]. 10 pp. Reprinted from *Atti del R. Istituto veneto di Scienze, Lettere, Arti,* Series 5, Vol. 6. [*see* p.].

2372 a d

— Sulle iscrizioni ed altri monumenti reto-euganei: dissertazione. Padova, Co' Tipi di Angelo Sicca, 1853. 48 pp., plates. [*see* p.].

2373 a d i

— Zodiaco etrusco, pietra euganea, ustino romano: tre notizie archeologiche. Padova, Co' Tipi di Angelo Sicca, 1856. 24 pp., plates. [*see* p.].

2374 a d

Schliemann, H. Mycenae: a narrative of researches and discoveries at Mycenae and Tiryns. London, John Murray, 1878. lxviii, 384 pp., front., illus., plates.

2375

Semmering. Südbahn-Hotel, Semmering: Klimatischer und Höhencurort. Wien, Verlag der K.K. Priv. Südbahn-Gesellschaft, [1882]. 9 pp., illus. At head of title: Prospect. [*see* p.].

2376

[Seymour, W. 2nd Duke of Somerset]. The Lord Marquesse of Hertford, his letter sent to the Queen in Holland. London, Joseph Hunscott & John Wright, 1642. 8, 7 pp.

2377 d i

Smythe, E. A., Viscountess Strangford. The eastern shores of the Adriatic in 1863, with a visit to Montenegro. London, Richard Bentley, 1864. [vii], 286 pp., plates. (to IB, 1872).

2378 a d

Stanford, E., publisher. An ethnological map of European Turkey & Greece, with introductory remarks on the distribution of races in the Illyrian peninsula, and statistical tables of population. London, Edward Stanford, [1877]. 32 pp., map, tables. [*see* p.].

2379 d i

Steinbuechel-Rheinwall, A. de. Un balletto di duemila trecento e più anni fa: bozzetto archeologico a dichiarazione di un passo di Erodoto. [Trieste, 1876]. 15 pp. Reprinted from *Archeografo triestino,* N.S., Vol. 4, No. 3. [*see* p.].

a

Another copy. [*see* p.].

2380

Stirling, D. McN. The beauties of Shute . . . and a succinct account of the illustrious family of Pole. Exeter, printed for the author, 1834. 24 pp. [*see* p.].

2381 d i

Stillman, W. J. Herzegovina and the late uprising: the causes of the latter and the remedies. London, Longmans, Green, 1877. iv, 186 pp. (Ventnor, 7 Feb.).

2382 e

Stossich, A. Una escursione botanica sul Monte Slavnik nel litorale. [n.p., n.d.]. 20 pp. Reprinted from an unidentified journal. [*see* p.].

2383 k

Stuart, G. The history of Scotland, from the establishment of the Reformation, till the death of Queen Mary; second edition. London, J. Murray, 1783–4. 2 vols.

2384 a

[Stuart, J., ed.]. Sculptured stones of Scotland. Vol. 2. Edinburgh, printed for the Spalding Club, 1867. lxxv–lxxxv pp., plates. [*see* q.].

2385 a l

Stuart-Glennie, J. S., ed. Greek folk-songs from the Ottoman provinces of northern

Hellas; literal and metrical translations by Lucy M. J. Garnett; classified, and edited, with essays on the survival of paganism, and the science of folk-lore; second edition, revised and enlarged. London, Ward & Downey, 1888. xxxix, 290 pp.

2386 a d
Taramelli, T. Escursioni geologiche fatte nell'anno 1872. [Udine, 1872]. 29 pp., diagr. Reprinted from *Annali scientifici del R. Istituto Tecno. di Udine,* Vol. 6. [*see* p.].

2387 a d i
Taylor, I. The Etruscan language. London, Robert Hardwicke, 1876. 24 pp., illus. Reprinted from the *Journal of the Transactions of the Victoria Institute, or Philosophical Society of Great Britain.* [*see* 2388].

2388 a d h i j
— Etruscan researches. London, Macmillan, 1874. xii, 388 pp., front., illus. (J. Gossard, Bologne, 25 Feb. 1885; Hyde Clark, London, 28 March 1874; London, 8 April 1874).

2389
Territet-Montreux-Glion. Le chemin de fer funiculaire à crémaillère: Territet-Montreux-Glion. Lausanne, Imprimerie Lucien Vincent, 1884. 24 pp., illus. [*see* p.].

2390 h i
Terstenjak, D. Slovanski elementi v Venetščini: spisal. v Ljubljani, Natisnili J. Blaznikovi, dediči, 1877. 91 pp. From *Letopisa Matica slovenska,* 1877. [*see* p.].

2391 a d
Thayer, A. W. L'avvenire commerciale di Trieste: rapporto A.S.E. Il Sig. John Jay, Ministro Plenipotenziario degli Stati Uniti d'America in Vienna, di Alexander W. Thayer, Console degli Stati Uniti in Trieste, 1869; traduzione per Cesare Combi. Trieste, Tipografia Peternelli & Morterra, 1872. 23 pp. [*see* p.].

2392 d i
Tommasini, M. Cenni storici e fisici sulla selvicoltura dell'agro triestino. Trieste, Tipografia del Lloyd Austro-Ungarico, 1876. 56 pp., plates. Reprinted from *Bollettino dell scienze naturali in Trieste,* Vol. 2, No. 1. [*see* p.].

2393
Trieste. Costituzione della città immediata di Trieste, 12 aprile 1850: istruzione per le elezioni del consiglio regolamento interno . . . Trieste, Stab. Art. Tip. G. Caprin, 1879. xiv, 221 pp.

2394 a d
Troll, U. von. Letters on Iceland: containing observations on the civil, literary, ecclesiastical and natural history . . . made, during a voyage undertaken in the year 1772 . . . second edition, corrected and improved. London, J. Robson, [etc.], 1780. xxiv, 400 pp., front. (map), plates.

2395 d
Turnbull, P. E. Austria. London, John Murray, 1840. 2 vols.

2396 a d
[Urbas, W.]. Die Slovenen: ethnographische Skizze. Triest, Buchdruckerei des Oesterreichischen Lloyd, 1873. 63 pp. From *Dritter Jahresbericht der deutschen Staats-Oberrealschule in Triest für das Schuljahr 1873.* [*see* p.].

2397 l
Urquhart, D. The spirit of the East, illustrated in a journal of travels through Roumeli during an eventful period. London, Henry Colburn, 1838. 2 vols., plate.

2398 d i
Valcourt, — de, and V. Petit. Cannes: son climat et des promenades; troisième édition. Paris, Germer Baillière; Cannes, Robaudy, 1878. 343 pp., front., diagrs.

2399
Vaux, W. S. W. Extracts from letters addressed to C. T. Newton, Esq., by M. Demetrius Pierides and F. Calvert, Esq. [London, 1863]. 5 pp., plate. Reprinted from the *Transactions of the Royal Society of Literature,* Series 2, Vol. 7. [*see* 241].

2400 e
— On a Greek inscription from Saloniki, Thessalonica. [London, 1866]. 24 pp., front., plates. Reprinted from the *Transactions of the Royal Society of Literature,* Series 2, Vol. 8. [*see* 241].

2401
Vienna. Vienne: ses monuments, musées, curiosités, environs; suivis d'une histoire abrégée de cette capitale; cinquième édition, revue et corrigée. Vienne, Charles Gerold Fils, 1873. [vi], 208 pp., map, plan.

2402 a d
Vojnović, C. Lettere dalla Croazia. Zara, Tipografia del Nazionale, 1873. 162 pp. [*see* p.].

2403 d k
[Voltaire, F. M. Arouet de]. Histoire de Charles XII, Roi de Suede, par Mr. V***; seconde édition, révue & corrigée par l'auteur. Basle, Christophe Revis, 1732. 8 pts. (in 1), front.
2404 k
— Le siècle de Louis XIV; nouvelle édition. Londres, 1788. 2 vols.
2405 i
Walford, E. The pilgrim at home; published under the direction of the Tract committee. London, Society for Promoting Christian Knowledge, [etc.]. 1886. 252 pp., front. (to IB).
2406 d
Walker, W. F. The Azores: or Western islands: a political, commercial and geographical account. London, Trübner, 1886. viii, 335 pp., front., illus., plates.
2407 d h j
Watts, W. L. Snioland: or, Iceland: its jökulls and fjallis. London, Longmans, 1875. 183 pp., front., plates, map. (W. L. Watts, Reyhirleith, 23 July 1875).
2408
Weiss, K. Der Dom zu Agram. Wien, K.K. Hof- und Staatsdruckerei, 1860. 38 pp., illus., plate. Reprinted from *Mittheilungen der k.k. Central-Commission zur Erforschung und Erhaltung der Baudenkmale,* Vol. 4. [*see* p.].
2409 d l
Wilkinson, Sir J. G. Dalmatia and Montenegro; with a journey to Mostar in Herzegovina and remarks on the Slavonic nations. London, John Murray, 1848. 2 vols., fronts., plates, tables.
2410 d h
Woerl's Reisenhandbuecher. Maria-Zell und Umgebung. Würzburg, Wien, Leo Woerl, [1887]. 28 pp. [*see* p.].
h l
Another copy. [*see* p.].
2411 i
Wright, T. The ruins of the Roman city of Uriconium, at Wroxeter, near Shrewsbury; fourth edition. Shrewsbury, J. O. Sandford, 1864. 101 pp., front., plates. (to IB).
Two other copies; third copy dated 1868, the 'fifth edition'.
2412 a d
Yriarte, C. La vie d'un patricien de Venise au seizième siècle. Paris, E. Plon, 1874. [iii], 447 pp., front., plate.
2413 e (p. 1)
— Trieste e l'Istria. Milano, Fratelli Treves, 1875. [vii], 131 pp., front., illus.
2414 e i
Yule, A. F. A little light on Cretan insurrection. London, John Murray, 1879. ix, 146 pp.
2415
Zandonati, V. Guida storica dell'antica Aquileja. Gorizia, Tipografia di G. B. Seitz, 1849. 232 pp., front.
2416 a d
Zannoni, A. Sugli scavi della Certosa relazione letta all'inaugurazione del Museo civico di Bologna, il 2 ottobre 1871 . . . Bologna, Regia Tipografia, 1871. 56 pp. [*see* 2149].
2417 a d
Zara. Programma dell' I. R. Ginnasio completo di prima classe in Zara alla fine dell' anno scolastico 1859–60. Zara, Tipografia Governiale, 1860. 2 pts. (in 1), tables.
2418
Zurich. Zurich guide; published by the Official General Inquiry Office. [Zurich, after 1888]. [vi], 26 pp., illus., map. [*see* p.].

2419-26 Pacific, Australasia

2419
[Bugnion, F. L.]. Bishop Bugnion's colony in Australia. Genève, Imprimerie Ramboz et Schuchardt, 1876. 15 pp. [*see* p.].

2420
[Cook, J.]. Captain Cook's relics. [London], 1887. [2] pp., illus. From *The Graphic,* 1 Oct. 1887. [*see* p.].

2421 a d i
Ellis, H. T. Hong Kong to Manila, and the lakes of Luzon, in the Philippine isles, in the year 1856. London, Smith, Elder, [etc.], 1859. viii, 294 pp., front., illus., plates, map. (9 Sept. 1872).

2422 a d i l
Fenton, F. D. Suggestions for a history of the origin and migrations of the Maori people. Auckland, N.Z., H. Brett, "Evening Star Office", 1885. [iv], 132 pp., front., table. (G. W. Rusden, 16 Nov. 1885).
i
Another copy (probably a proof copy) bound in front of the first copy above.

2423 a d
Kelly, W. Life in Victoria: or, Victoria in 1853, and Victoria in 1858, showing the march of improvement made by the Colony within those periods, in town and country, cities and diggings. London, Chapman & Hall, [etc.], 1859. 2 vols., illus., map.

2424 d
Paul, R. B. New Zealand, as it was and as it is. London, Edward Stanford, 1861. [iii], 64 pp. [*see* 223].

2425
Sinnett, F. An account of the Colony of South Australia; prepared for distribution at the International Exhibition of 1862. London, Robert K. Burt, [1862]. [iv], 96 pp., front. (map). [*see* 240].

2426 a d
Smyth, R. B. The gold fields and mineral districts of Victoria, with notes on the modes of occurrence of gold and other metals and minerals. Melbourne, John Ferres; London, Trübner, 1869. vi, 644 pp., front., illus., plates, maps.

2427–536 Maps

General

2427 a d

Atlas of classical geography completed to the present state of knowledge . . . London, H. G. Bohn, 1861. v, 24 pp., 22 maps. (Standard library).

2428

Evans, Sir F. J. O. Chart of the curves of equal magnetic variation, 1858. London, Admiralty, 1858. 1 sheet. [*see* p.].

2429 d

Johnston, A. K. The royal atlas of modern geography; new edition. Edinburgh, London, W. & A. K. Johnston, 1878. xi, [152] pp., front. (map), 50 maps. [*see* q.].

2430 a d l

Ptolemy, Claudius. Claudii Ptolemaei Alexandrini geographicae enarrationis libri octo. ex Bilibaldi Pirckeymheri tralatione . . . Lugduni, ex Officina Melchioris et Gasparis Trechsel Fratrum, 1535. 150, [73] pp., 50 plates. [*see* q.].

2431 d k

Society for the Diffusion of Useful Knowledge. The cyclopaedian: or atlas of general maps . . . [and] Index to the principal places in the world with reference to the maps, by James Mickleburgh. London, Edward Stanford, 1861. [iii] pp., 39 maps; [i], 70 pp. [*see* q.].

2432 l

— The Harrow school atlas of classical geography. London, Edward Stanford, [*c.* 1860]. [iii] pp., 23 maps; 48 pp. [*see* q.].

Maps – *Africa*

2433 l

Algeria. Carte des chemins de fer de l'Algérie et de la Tunisie [and of southern Europe. Paris, 1868?]. 1 sheet. [*see* p.].

2434

Arlett, W. Gran Canaria surveyed . . . 1834. London, Admiralty, Hydrographic Office, 1848. 1 sheet. At head of title: Canary Islands. [*see* cabinet].

2435 i

Baumann, O. Karte der Insel Fernando Póo, nach eigenen Aufnahmen konstruiert und gezeichnet . . . mitglied der Prof. Dr. Lenz'schen Expedition nach Aequatorialafrika 1885–87. Gotha, Justus Perthes, 1887. 1 sheet. (Vienna, 21 Nov. 1887). [*see* p.].

2436

Church Missionary Society. Imperfect sketch of a map from 1½° north, to 10½° south latitude, and from 29 to 44 degrees east longitude, by the missionaries of the Church Missiony. Society in eastern Africa. Rabbai Mpia, J. Rebman, 1850. 1 sheet. [*see* p.].

2437 d f

Compagnie Universelle du Canal maritime de Suez. Carte de l'isthme. Paris, E. Andriveau-Goujon, 1869. 1 sheet. [*see* p.].

2438

Dolben, W. D. M. and T. Stringer. Africa – West coast: river Volta sketch survey . . . 1861; the coast line & outer soundings from the Admiralty coast chart. London, Admiralty, 1862. 1 sheet. [*see* p.].

2439
Graphic. Supplement. Birdseye view of Egypt. London, 1882. 1 sheet. (Supplement to the *Graphic,* 12 Aug. 1882). [*see* cabinet].
2440
Gross, R. Südliches Wolta-Gebiet auf der Goldküste von West-Afrika, nach den Angaben der Missionare Locher & Plessing. [n.p.], Rudolf Gross, [after 1855]. 1 sheet. [*see* p.].
2441
Hassenstein, B. Dr. W. Junker's Reisen in Nordost-u. Central-Afrika. Blatt No. 2: Karte der Routen in den Mudiríen, Rohl u. Bahr-el-Ghasál, sowie Übersicht der wichtigsten neueren Reisen in den Ägyptischen Aquatorialprovinzen. Gotha, Justus Perthes, 1880. 1 sheet. From *Petermann's geographische Mitteilungen,* Jahrgang 1880, Tafel 4. [*see* cabinet].
2442
Homem, D. Africa, extrahido do atlas m.s. feito . . . em 1558 existente no Museo britannico. London, Edward Stanford, 1860. 3 sheets. Sheet 3 entitled: Continente africano.
2443
Johnston, K. Africa. Edinburgh, London, William Blackwood, [1879?]. 1 sheet. [*see* p.].
2444
Macdonald, J. R. L. Map of Uganda and adjoining territories compiled for the Intelligence Division, War Office. London, Edward Stanford, 1895. 1 sheet. [*see* cabinet].
2445
Mahmoud. Carte de l'antique Alexandrie et de ses faubourgs . . . fait en 1866. Paris, Imp. Monrocq, 1866. 1 sheet. [*see* p.].
2446
Medlycott, M. B. and T. H. Flood. Plan of creeks of river Congo. [London, *c.* 1870]. 1 sheet. [*see* cabinet].
2447 d
Owen, W. F. W. The east coast of Africa, including the islands of Zanzibar & Pemba, by Capt. A. T. E. Vidal & the officers . . . under the orders of Captn. W. F. W. Owen, 1823–4. London, Admiralty, Hydrographical Office, 1828. 1 sheet. [*see* cabinet].
2448 d
— Island and ports of Mombaza surveyed by Lieutts. W. Mudge, T. Boteler, & R. Owen . . . under the directions of Captn. W. F. W. Owen of H.M.S. Leven. London, Admiralty, Hydrographical Office, 1827. 1 sheet. [*see* cabinet].
2449
— West coast of Africa . . . Fernando Po to Cape Lopez, surveyed by Captain W. F. W. Owen, R.N., 1826, and Captain A. T. E. Vidal and Lieutenant Bedford, 1836–1838. London, Admiralty, Hydrographic Office, 1841. 1 sheet. [*see* cabinet].
2450
Prout, G. H. Carte du Kordofan dressée . . . [Cairo], 1876. 1 sheet. From *Bulletin de la Société khédiviale de Géographie,* Series 2, Vol. 5. [*see* p.].
2451 d
Ravenstein, E. G. Afrika, nordwestl. Blatt; Afrika nordöstl. Blatt. Hildburghausen, Verlag des Bibliographischen Instituts, 1863. 1 sheet. [*see* p.].
2452 d
— Afrika, südliches Blatt. Hildburghausen, Verlag des Bibliographischen Instituts, 1862. 1 sheet. [*see* p.].
2453 d
Royal Navy and Merchant Service. A chart . . . of the coast of Africa comprised between the Isles de Loss and the Bight of Biafra, including the Windward and Gold Coasts . . . London, Richd. H. Laurie, chartseller to the Admiralty, 1859. 1 sheet. [*see* p.].
2454 d
Vidal, A. T. E. and W. Arlett. The currents of the coast of Guinea collected . . . 1840. [London], Admiralty, Hydrographic Office, 1847. 1 sheet. [*see* p.].
2455
— Tenerife, surveyed . . . 1838. [London], Admiralty, Hydrographic Office, 1848. 1 sheet. At head of title: Canary Islands. [*see* cabinet].
2456
— The Canary Islands, surveyed . . . 1834 & 1838. [London], Admiralty, Hydrographic Office, 1848. 1 sheet. [*see* cabinet].

2457
— and G. A. Bedford. Africa – West coast: Republic of Liberia, formerly called the Grain coast, the coast line & soundings. London, Admiralty, 1862. 1 sheet. [*see* p.].

Maps – *Americas*

2458 d
Arrowsmith, J. The provinces of the Rio de la Plata and adjacent countries, chiefly from the map constructed . . . for Sir Woodbine Parish . . . drawn by Augustus Petermann. [London], A. Petermann, [1850?]. 1 sheet. [*see* p.].

2459
Atacama. Provincia de Atacama. [Santiago?, 18—]. 1 sheet. [*see* cabinet].

2460
Brazil. [Mail route from São Paulo. n.p., 18—]. 1 sheet (holograph). [*see* p.].

2461 d
Burmeister, H. Karte zur übersicht von H. Burmeister's Reise in Brasilien im Jahre 1850. Berlin, Geo. lithograph. Anst. v. H. Mahlmann, 1850. 1 sheet. Upper cover: Provincia do Rio de Janeiro e Minas. [*see* p.].

2462
[Campos do Jacuhy e Taquary]. Lipsia, Instituto lithographico de H. Kunsch, 1861. 1 sheet. [*see* cabinet].

2463
Colton, J. H., publisher. Colton's map of the United States, the Canadas &c. showing the rail roads, canals & stage roads . . . New York, J. H. Colton, 1860. 1 sheet. [*see* p.].

2464 d
Gerber, H. Carta da Provincia de Minas Geraes; coordeuado por ordem do Exm. Sr. Conselheiro José Bento da Cunha Figueiredo, Presidente do Provincia . . . [Rio de Janeiro], 1862. 1 sheet. [*see* p.].

2465 i
Harrison, T. Map of the Isthmus of Panama representing the line of the Panama rail road as constructed under the direction of George M. Totten . . . Jamaica, 1857. New York, Endicott, 1857. 1 sheet. (W. Nelson). [*see* p.].

2466 d h
Johnston, A. K. Stanford's library map of South America. London, Edward Stanford, [1864]. 3 sheets. [*see* p.].

2467 d
Laberge, A. M. Plano de la ciudad de Cordoba. Paris, Imp. Destouches, 1860. 1 sheet. [*see* p.].

2468
Laemmert, E. & H., publisher. Nova planta de cidade do Rio de Janeiro. [Rio de Janeiro], 1864. 1 sheet. [*see* p.].

2469 d
Liais, E. Hydrographie du haut San-Francisco et du Rio das Velhas ou résultats au point de vue hydrographique d'un voyage effectué dans la Province de Minas-Geraes. Paris, Garnier Frères; Rio de Janeiro, B. L. Garnier, 1865. iii, 27 pp., 19 maps. At head of title: Explorations scientifiques au Brésil. [*see* q.].

2470 a d h
Mendes de Almeida, C. Atlas do Imperio do Brazil comprehendendo as respectivas divisões administrativas, ecclesiasticas, eleitoraes e judiciarias destinado á dos alumnos do Imperial Collegio de Pedro II. Rio de Janeiro, Lithographia do Instituto philomathico, 1868. 37 pp., 34 maps.

2471 d
Mouchez, E. Carte routière de la côte du Brésil de Rio Janeiro au Rio de la Plata et au Paraquay. [Rio de Janeiro], Dépôt des Cartes et Plans de la Marine, 1864. 1 sheet. [*see* cabinet].

2472 l
Moussy, V. M. de. Carte de la Confédération Argentine . . . et des pays voisins, État oriental de l'Uruguay, Paraquay, partie du Brésil et de la Bolivie. Paris, Imp. Lemercier, 1867. 1 sheet. [*see* p.].

2473 l

— Carte de la Province de Corrientes du territoire des missions et des pays adjacents. Paris, Imp. Lemercier, 1865. 1 sheet. [*see* cabinet].

2474 d i

Mueller, D. P. Mappa chorographico da Provincia de São Paulo . . . Paris, Alexis Orgiazzi, 1837. 1 sheet. (A. Moreira de Barros, Taubate, 14 Feb. 1866). [*see* p.].

2475

Niemeyer, C. J. de. Nova carta chorographica do Imperio do Brazil . . . 1856. [Rio de Janeiro], Gravada na Litha. do Archo. Militar por Alvaro e Pereira, 1867. 1 sheet. [*see* p.].

2476

Sao Paulo. Mappa da costa da Provincia de S. Paulo, 5 de abril de 1867. 1 sheet (manuscript). [*see* cabinet].

2477

Schultz, W. Mappa da Provincia de Santa Catharina do Imperio do Brasil com as partes adjacentes das Provincias do Parana e de São Pedro do Rio Grande do Sul. Leipzig, Instituto lithographico de F. A. Brockhaus, 1863. 1 sheet. [*see* cabinet].

2478

— Planta das colonias allemães e terrenos medidos na Provincia de Santa Catherina. Lipsia, Instituto lithographico de H. Kunsch, 1861. 1 sheet. [*see* cabinet].

2479 d l

Sourdeaux, A. Plano topografico de los alrededores de Buenos Ayres levantado con licencia del Superior Gobierno. [Buenos Aires], Libreria de la Victoria, [*c.* 186-]. 1 sheet. [*see* cabinet].

2480 l

Stanford, Edward, publisher. Stanford's map of the seat of war in America. London, Edward Stanford, 1861. 1 sheet. [*see* p.].

2481

United States of America. United States geological exploration of the fortieth parallel. Atlas accompanying Volume III on mining industry . . . compiled and drawn by R. H. Stretch, from the Mine Surveys of I. E. James and R. H. Stretch, 1863 to 1869, Marlette and Hunt, 1864 to 1868 and T. D. Parkinson, 1869. [Washington, D.C.], New York, Julius Bien, [1871]. 14 maps. [*see* cabinet].

Maps – *Asia*

2482 l

Arabia. Map of south western Arabia. [London], War Office, Topographical Dépôt, [1874]. 1 sheet. [*see* p.].

2483

Blondel, L. A. Carte du Liban d'après les reconnaissances de la Brigade topographique du Corps expéditionnaire de Syrie en 1860–1861, dressée au Dépôt de la Guerre . . . Paris, Imp. Lemercier, 1862. 1 sheet. [*see* p.].

2484 d

Bombay with the harbour and country adjacent, from the best authorities. London, J. M. Richardson, lithographer, [18—]. 1 sheet. [*see* cabinet].

2485

Bond, H. and F. W. Sullivan. Sketch of Saida (ancient Sidon) . . . 1860. London, Admiralty, 1861. 1 sheet. [*see* cabinet].

2486

Dillon, C. H. Beïrout bay, surveyed . . . 1842. London, Admiralty, Hydrographic Office, 1844. 1 sheet. At head of title: Syria. [*see* cabinet].

2487 l

Goa. The Goa's map gen[er]al transcribed. [n.p., n.d.]. 1 sheet (manuscript). [*see* cabinet].

2488

Gonsalves, J. M. Mappa do territorio portuguez de Goa . . . extrahido d'original . . . Bombaim, [187-]. 1 sheet. [*see* p.].

2489

Great Britain. War Office. Intelligence Division. Map of the Pamirs and adjacent territory. London, Edward Stanford for the War Office, 1893. 1 sheet. [*see* cabinet].

2490 d l

Hughes, W. Arabia, the Red Sea, and the valley of the Nile, including Egypt, Nubia, and Abyssinia. London, Liverpool, George Philip, [1867]. 1 sheet. [*see* p.].

2491

India. Survey of India Dept. Agra cantonment, city and environs: season 1868–69. [Calcutta, 1870]. 6 sheets. [*see* cabinet].

2492

— — Contour map of India. [Calcutta, 1886]. 6 sheets. [*see* cabinet].

2493

— — Railway map of India. [Calcutta, 1885]. 6 sheets. [*see* cabinet].

2494

Kiepert, H. Karte von Armenien, Kurdistan und Azerbeidschan . . . entworfen und bearbeitet 1852–55. Berlin, Simon Schropp, 1858. 1 sheet. [*see* p.].

2495

Mansell, A. L. Beirút, the antient Berytus, surveyed . . . 1859. London, Admiralty, 1860. 1 sheet. [*see* cabinet].

2496

Saunders, T., compiler. Surveys of ancient Babylon and the surrounding ruins . . . made by order of the Government of India in 1860 to 1865 . . . [by W. B. Selby, W. Collingwood and J. B. Bewsher]. London, W. H. Allen, Edward Stanford, Trübner, 1885. 6 sheets. [*see* cabinet].

2497

Selby, W. B. This plan of the supposed ruins of Babylon . . . assisted by . . . W. Collingwood . . . [London?], J. & C. Walker, engraver, [1874?]. 1 sheet. [*see* cabinet]. At head of title: His Highness Omar Pasha . . . Generalissimo of the Ottoman Armies, and Governor General of the Turkish Arabia.

2498

Velde, C. W. M. van de. Map of the Holy Land; second edition. Gotha, Justus Perthes, 1865. 1 sheet. [*see* p.].

2499

Wallin, G. A. Map of N. west Arabia. [London, 1850]. 1 sheet (holograph). From the *Journal of the Royal Geographical Society,* Vol. 20. [*see* p.].

2500

Wilson, C. W. Ordnance survey of Jerusalem . . . under the direction of Colonel Sir Henry James . . . 1864–5. Southampton, Ordnance Survey Office, 1876. 1 sheet. [*see* cabinet].

Maps – *Europe*

2501

Adriatisches Meer. [Trieste?, 187–]. 1 sheet. [*see* cabinet].

2502 l

Andriveau-Goujon, E., publisher. Chemins de fer de l'Europe centrale. Paris, 1884. 1 sheet. [*see* p.].

2503

— — Navigation à vapeur dans le bassin de la Méditerranée et chemins de fer de l'Europe centrale. Paris, E. Andriveau-Goujou; London, Edward Stanford, 1868. 1 sheet. [*see* p.].

2504 d

Aquileia. Ichnographia Aquileiae Romanae et Patriarchalis: piano topografico d'Aquileja. [Vienna], Lith. u. ged. i.d. K.K. Hof- u. Staatsdruckerei, [18—]. 1 sheet. [*see* p.].

2505

Austria. [Southern Austria and part of the gulf of Venezia. Vienna?], C. Stein, 1875. 1 sheet. [*see* p.].

2506 a

Bossoli, E. F. Panorama des Alpes . . . vu du Grand Hôtel Varese, dessiné d'après nature. Turin, Lit. F. Doyen, 1876. [*see* p.].

2507 d

Brindisi. Piano generale del porto di Brindisi, rilevato del 1866. [Florence?], L. Balatri Autografó, 1866. 1 sheet. [*see* p.].

2508
Chatoff, — General. Karte der Walachei, Bulgarien und Rumelien. Berlin, S. Schropp, [*c.* 1854]. 1 sheet. [*see* p.].
2509
Club Alpino Austro-Germanico. Sezione di Trieste. Panorama delle Alpi carniche e goriziane preso da Opcina presso Trieste. Vienna, Stamperia di Corte ed Instituto artistico di G. Reiffenstein, [18—]. 1 sheet. [*see* cabinet].
2510
Great Britain. Admiralty. English channel, the coast of England and Ireland from various surveys; the coast of France from the Pilote français. London, Admiralty, 1872. 1 sheet. [*see* cabinet].
2511
— — Gibraltar bay from the Spanish survey of 1872. London, Admiralty, 1877. 1 sheet. At head of title: Strait of Gibraltar. [*see* cabinet].
2512 l
— — Iceland and the Faeroe islands from the Danish survey 1845 corrected to 1866 . . . London, Admiralty, [1872]. 1 sheet. [*see* p.]. Another copy.
2513
— — Kronstadt north and south channels from Russian surveys to 1867. London, Admiralty, 1871. 1 sheet. At head of title: Gulf of Finland. [*see* cabinet].
2514
Greenwood, C. Map of the County Palatine of Lancaster, from an actual survey, made in the year 1818. London, Greenwood, Pringle, 1834. 1 sheet. [*see* p.].

2515 l
Greenwood, Pringle, publisher. [Map of south Lancashire] to the nobility, clergy & gentry of Lancashire this map is respectfully dedicated by the proprietors. [London, 1838?]. 1 sheet. [*see* p.].
2516
Istria. Milano, Stabilimento di L. Ronchi, [1878?]. 1 sheet. [*see* cabinet].
2517
Kiepert, H. New original map of the island of Cyprus. Berlin, Dietrich Reimer; London, Williams & Norgate, 1878. 1 sheet. [*see* p.].
2518 d
Laibach. Umgebungen von Laibach und Adelsberg. [Vienna?], D. Huber, C. Stein, 1871. 1 sheet. [*see* cabinet].
2519 d
Marinelli, G. and T. Taramelli. Carta del Friuli tra i Fiumi Livenza ed Isonzo. Udine, Lit. E. Passero, 1879. 1 sheet. [*see* cabinet].
2520 d h l
Mayr, G. Atlas der Alpenländer, Blatt VI, südl. Steyermark, Illirien, (Kärnthen, Krain) Friaul, Küstenland &c. Gotha, Justus Perthes, 1870. 1 sheet. [*see* p.].
2521
Militaerisch-geograpisches Institut in Wien. [Euboea]. Wien, 1884. 1 sheet. [*see* p.].

2522
— Special-Karte des Koenigreiches Dalmatien, astronomisch-trigonometrisch Vermessen, topographisch Aufgenommen, reduzirt, gezeichnet und gestochen . . . herausgegeben in den Jahren 1861 bis 1863. [Vienna], 1873. 6 parts. [*see* q.].
2523
Monaco. Carte de la haute Italie littoral de la Méditerranée aux environs de Monaco, Menton, Sn. Rémo, Port Maurice, Savone, Voltri et Gênes. Cannes, F. Robaudy, 1880. 1 sheet. [*see* cabinet].
2524
Murchison, Sir R. I. Geological map of England and Wales . . . according to the most recent researches; fourth edition. London, Edward Stanford, 1858. 1 sheet. [*see* cabinet].
2525 d
Ólsen, O. N. Island, efter Oluf Nicolai Ólsen's og Björn Gunnlaugssons Uppdrattr Íslands. Copenhagen, Boerentzen, 1849. 1 sheet. [*see* p.].
2526 d
— Uppdráttr Íslands, gjördr ad fyrirsögn Ólafs Nikolas Ólsens eptir landmaelíngum Bjarnar Gunnlaugssonar. Reykjavík, Kaupmannahöfn, Grafid hefir F. C. Holm, 1844. 1 sheet. [*see* p.].

2527
Pfeiffer, F. Scheletro stradale della città di Trieste. Trieste, Julius Dase, [1867?]. 1 sheet. [*see* p.].
2528
Philip & Son, George, publisher. Philip's new map of central Europe. London, George Philip, [1866?]. 1 sheet. (Philips' series of maps for travellers). [*see* p.].
2529
Railway Commissioners. A general map of Ireland to accompany the report of the Railway Commissioners shewing the principal physical features and geological structure of the country constructed in 1836 and engraved in 1837-8. Dublin, Hodges & Smith; London, James Gardner, [1838]. 6 sheets. [*see* q.].
2530
Smith & Son's, W. H., publisher. New railway map of the British Isles. London, [etc., 1865?]. 1 sheet. [*see* p.].
2531
Switzerland. [Map. London], J. Walker, [18—]. 1 sheet. [*see* cabinet].
2532
Trieste. Erhöhung über dem Meere in Wiener Klafter der trigonometrisch bestimmten vorzüglichsten Punkte in der spezial Karte von Steyermark und dem vormaligen Königreiche Illyrien. [Vienna?], F. von Schönfelder, C. Stein, 1871. 1 sheet. [*see* p.].
2533
— Nuova rete ferroviaria per avvicinare Trieste alla Pontebba, al Brennero ed all'Italia. Trieste, H. Rieter, [18—]. 1 sheet. [*see* p.].
2534
Waterlow & Sons Ltd., publisher. Map of the continental railways in connection with the London, Brighton & south coast and western of France railways. London, [18—]. 1 sheet. [*see* p.].

Maps – *Pacific*

2535 d h
Pacific ocean. [London], 1864. 1 sheet. [*see* p.].
2536
Sterndale, H. B. Karte des Landes zwischen den Flüssen Sigaco und Letoga sowie der Ansiedelungen am Hafen von Apia . . . Archipel der Samoan . . . 1870; bearbeitet und gezeichnet von L. Friederichsen. Hamburg, L. Friederischen, 1873. 1 sheet. From *Journal des Museum Godeffrey,* No. 2, plate 1. [*see* p.].

2537-52 Autograph Letters (excluding those inserted in books) and other Manuscripts

2537
Letter to IB, signed: N. MacColl, [London], 10 March 1885. 1 leaf.
2538
Letter to Burton, signed: L. J. Maclagen, [London], 9 Aug. 1887. 1 leaf.
2539
Letter to Miss H. E. Bishop, signed: Kate Wright, Lincoln, 9 Oct. 1888. [4] pp.; sketches of swords, and cutting inserted (*see The Life,* Vol. 2, pp. 189, 253, 275).
2540
Anonymous. [History of the Funj dynasty of Nubia and the Sudan, from the foundation of the dynasty by 'Amára in 915 A.H. (1509 A.D.); Arabic text]. Dated in colophon 1295 A.H. [1878 A.D.]. 43 pp.
2541
— [Fan-English vocabulary]. 24 leaves. Annotated by Burton.
2542 h j
— [List of books on swords and fencing, mainly published in Spain]. 54 leaves. (Fred. W. Foster, [London], 25 Nov. 1884).
2543
— [Prescription note-book]. 26 pp. Pp. 23-5 notes by Burton.
2544 h
— [Transliteration & translation of the Great Nebuchad Nezzar inscription. Vol. 1]. ii, 62 leaves.
2545
Arabian nights. [Stories from the Arabian nights; Arabic text]. 17 leaves.
2546 d j
— [Photograph of the manuscript of the Arabian nights; Arabic text]. Vols. 3–7 (in 4). (H. W. Chandler, Oxford, 27 Aug. 1887). The manuscript in 7 vols., written between 1764–5, is in the Bodleian Library (MS. Bodl. Or. 550–6); it was purchased from Jonathan Scott [*see* 681]. *See also Supplemental nights,* Vol. 4, pp. 355-66 [*see* 94*a*].
2547 d g j l
— [Zayn al Asnàn, and Aladdin; Arabic text]. 306 leaves. Copy of the manuscript in the Bibliothèque Nationale, Paris sent by Professor Hermann Zotenberg, Keeper of the Oriental Manuscripts. (H. Zotenberg, Paris, 8 Feb. 1887 and 3 April 1888). *See also The Life,* Vol. 2, p. 333.
2548 l
Háfiz, Shírází. [Ghazals, or Persian poems of Háfiz; Persian text]. 326 pp. The manuscript belonged to Semuil-li-Arzani, a friend of Adam Clark, in 1810 A.D.
2549
Ross, W. A. On pyrology or fire chemistry. London, 1872. [ii] pp., 90 leaves.
2550 i
Sa'dí, Shírází. [Gulistán; Persian text]. 187 pp. (J. Lentaigne, Dublin Castle, 22 Aug. 1870). *See also* 95.
2551
Shiháb al-Dín Ahmad ibn 'Abd al-Qádir. [A manuscript copy of Shiháb al-Dín's account of the conquest of Abyssinia by Ahmad Grân in the sixteenth century, known as 'Fath al-habashah'; the text stops about 20 lines short of the text printed in Paris in 1897, *Histoire de la conquête de l'Abyssinie (XVIe siécle),* texte arabe, par René Basset, Vol. 1; *see* p. v for reference to Burton]. [203] leaves. [*see* q.].

2552

Turnbull, H. T. L. [On the processes which have taken place during the formation of the volcanic rocks of Iceland – On the intimate connection between the pseudo-volcanic phenomena of Iceland. 1861]. 2 vols. [leaves 1–116, 1–69, 70–82].

INDEX

*　　*　　*